# Spark
# 大数据实时分析实战

付　雯　聂　强　主　编

武春岭　李俊翰
王宁忆　李清莲　副主编
谭博文　郑　伟

文红亚　参　编

北京理工大学出版社
BEIJING INSTITUTE OF TECHNOLOGY PRESS

## 内 容 简 介

本书分为六个项目，通过真实大数据实时分析项目的导入，引导读者完成大数据实时分析平台Spark的搭建，通过对基于Hadoop生态圈中Yarn资源调度框架，搭建Spark日志管理系统，搭建Kafka分布式消息系统，在工作中实现使用SparkStreaming实时读取Kafka中的数据进行实时处理。

本书适用于大数据技术与应用、数据科学与大数据技术等电子信息类专业教学的教材，也可作为工程技术人员的参考书。



**图书在版编目（CIP）数据**

Spark大数据实时分析实战/付雯，聂强主编．—北京：北京理工大学出版社，2020.12（2023.7重印）

ISBN 978－7－5682－8844－6

Ⅰ.①S…　Ⅱ.①付…②聂…　Ⅲ.①数据处理软件　Ⅳ.①TP274

中国版本图书馆CIP数据核字（2020）第140447号

出版发行／北京理工大学出版社有限责任公司
社　　址／北京市海淀区中关村南大街5号
邮　　编／100081
电　　话／（010）68914775（总编室）
　　　　　（010）82562903（教材售后服务热线）
　　　　　（010）68948351（其他图书服务热线）
网　　址／http：//www.bitpress.com.cn
经　　销／全国各地新华书店
印　　刷／保定市中画美凯印刷有限公司
开　　本／787毫米×1092毫米　1/16
印　　张／16.25
字　　数／375千字
版　　次／2020年12月第1版　2023年7月第4次印刷
定　　价／75.00元

责任编辑／王玲玲
文案编辑／王玲玲
责任校对／周瑞红
责任印制／施胜娟

# Foreword 前言

近年来，智能设备和智能应用迅猛发展，极大地方便了人们的工作和日常生活，同时，也产生了大量的数据。这些应用和服务的成功得益于大数据和日益完善的实时分析技术。大数据实时分析技术的出现，可以对这些数据进行快速的分析，让决策者能够迅速地把握用户的关键需求，并能够及时响应用户的需求变化。未来几年，将有数以亿计的智能设备接入互联网，从智能手机、台式机、汽车到智能家居，都会接入互联网。未来的数据分析将对实时性要求越来越高。

针对大数据的迅猛发展，本书结合实际应用案例，选用高性能的 Spark 技术作为大数据实时分析的工具，介绍了实时大数据分析的实现过程，为读者提供了快速安装、搭建大数据分析集群和进行实时数据分析所需的技术。

本书采用项目驱动的编写方式，精心设计了 6 个项目，覆盖了 Spark 大数据实时分析技术所涉及的基础技术介绍、集群搭建，以及日志服务器搭建等知识技能点。书中深入浅出地介绍了 Spark 技术的基本原理、集群的构建与安装、基于 Yarn 的 Spark 集群搭建、Spark 日志服务器的搭建、Kafka 集群的构建与安装，并通过案例对 Kafka 集群进行了测试。具体内容为：

项目一从数据存储技术、分析技术、批数据和实时数据等数据分析背景知识入手，介绍了实时数据处理的价值、Spark 实时数据分析技术。

项目二通过具体的案例操作，详细介绍了 Spark 集群规划、安装准备、集群搭建、验证及客户端应用的搭建。

项目三通过案例介绍了基于 Yarn 的 Spark 集群的搭建、配置和验证的方法。

项目四介绍了 Spark 日志服务器的配置及验证日志服务器的方法。

项目五和项目六通过案例介绍了集群搭建、集群规划、安装配置，并对 Kafka 集群进行了测试。

本书由重庆电子工程职业学院付雯、聂强担任主编，重庆电子工程职业学院武春岭、李俊翰、王宁忆、李清莲、谭博文及潍坊职业学院郑伟担任副主编，来自重庆课外岛科技发展有限公司的文红亚高级工程师也参与了本书的部分编写工作。

尽管我们尽了最大努力，但书中难免有不妥之处，欢迎各界专家和读者朋友们给予宝贵意见和建议。

编　者

# Contents 目录

# 项目一
# 初识 Spark 技术

【项目描述】

对采集到的数据进行分析处理，可以获取有价值的信息。那么你知道数据分为哪些类型吗？不同的数据采用什么技术进行分析处理呢？

【项目分析】

近年来，大数据分析一词成了 IT 行业的流行术语，大数据分析即将大量的数据通过各种技术进行交叉分析，从而挖掘出数据背后的价值，甚至可以预见未来。要想通过大量的数据分析出数据的价值，关键在于收集数据、存储数据及分析数据。

## 任务 1.1　数据处理系统 

### 1.1.1　数据存储技术

在数据处理早期，互联网企业很少，公司数据量也不大，数据存储和处理领域还是关系型数据库（RDBMS）的天下，传统的关系型数据库的思想是过去 30 年形成的，致使关系型数据脱颖而出的是 Oracle 数据库，全世界的数据库市场几乎被 Oracle、IBM 的 DB2、Microsoft 的 SQL Server 垄断，其他数据库厂商市场份额比较小，后期又出现了一些开源的数据库，例如 MySQL、PostgreSQL。

这些数据库主要是面向 OLTP（on－line transaction processing，联机事务处理）需求设计开发的，以人机会话应用为主。以上这些数据库底层设计存储格式都是行存储，比较适合面向业务、数据频繁的增删改操作，但对于统计类的分析查询，行存储效率低下，当数据量达到千万级别量级时，复杂的 SQL 业务查询明显效率低下。

随着时间推移，互联网时代的到来，尤其是移动互联网的出现和发展，数据来源不再像传统数据一样，而是通过设备、服务器、应用等自动产生，例如手机、平板电脑终端产生的数据，以及大型集群收集的用户日志信息等。同时，传统行业数据量也在增长，互联网的数据样式多样，呈现结构化、半结构化及非结构化展现。总之，互联网时代的数据目前呈现出指数级别增长，这些数据涵盖各行各业，例如基因数据、教育数据、导航数据、卫星图片数据、气象数据、医疗数据等。

所谓的大数据应用，就是针对上述数据进行关联、交叉分析、比对，对数据进行深层次挖掘，例如，对数据进行即席查询、从非结构数据中提取关键信息、对半结构化的数据进行内容检索及复杂的计算分析等。针对以上数据业务要求的处理，传统的关系型数据库无论是在存储还是在业务分析功能上，都已经束手无策，这样的环境下就促使了类似 Hadoop 大数据技术的诞生，例如 Hadoop 生态圈中的 HBase NoSQL 分布式数据库、Hadoop 生态圈中的 Hive 数据仓库。

针对互联网大数据时代的数据处理压力，后期出现了很多类似于 Hadoop 生态圈中 HBase 的存储技术，这些技术都有一些共性：硬件基于服务器，存储基于服务器自身的磁盘，操作系统主要是 Linux，架构是基于大规模分布式计算且具有极高的横向扩展能力，除此之外，还提供故障容错及数据高可用保证机制。总体来看，大数据处理技术进入了新的阶段，这里的主要原因是传统的数据库技术及分析能力不能满足当前数据存储分析需求。目前大数据存储技术路线最典型的有如下三种。

**1. 大数据一体机**

这种是专为大数据存储分析处理而设计的软件、硬件结合的产品，由一组集成的服务器、存储设备、操作系统、数据库管理系统组成，同时，内部集成了为数据查询、处理分析预先安装及优化的软件。大数据一体机有良好的稳定性和纵向扩展能力。

**2. 采用 MPP 架构的数据库集群**

MPP（Massively Parallel Processing，大规模并行处理）数据库集群重点是面向行业大数据，采用 Shared Nothing 架构，通过列式存储、粗粒度索引，并结合大规模并行处理的高效的计算模式，完成对数据的分析。这类集群主要运行在低成本的 PC 机上，具有高性能和可扩展的特点，主要是针对结构化的数据进行存储和分析，相对于传统数据库技术来说，可以支撑 PB 级别的数据分析。

**3. 基于 Hadoop 生态圈技术扩展和封装**

这类技术主要是基于 Hadoop 生态技术衍生出来的，主要应对传统关系型数据库较难处理的数据和场景，例如对非结构化数据的存储及分析。目前最典型的应用场景就是通过扩展和封装 Hadoop 来实现大数据的存储分析，这类技术比较适用于非结构化数据处理、半结构化数据处理、复杂的 ETL 流程处理、数据挖掘等场景。

### 1.1.2 数据分析技术

目前，大数据领域每年都会产生很多新的技术，这些技术为大数据获取、存储、处理分析或可视化提供了有效的手段。其中，大数据分析技术能够将大规模数据中隐藏的价值信息挖掘出来，为社会经济活动提供依据，提高各个领域的运行效率。

大数据分析处理的基本流程和传统的数据处理流程的主要差别在于：大数据要处理大量的、非结构化的数据。所以，为了保证数据处理的速度，需要在并行的分布式系统中处理数据。目前，Hadoop、MapReduce 和 Spark 等分布式处理方式已经成为大数据处理过程中普遍使用的技术。

Hadoop 生态圈中提供了很多解决大数据问题的组件，如图 1 - 1 所示，这些组件包括数据存储、数据集成、数据处理和数据分析。例如，Hadoop 生态圈中的 HDFS 分布式存储系统是一个数据管理系统，其作为数据分析的源头，汇集了结构化和非结构化的数据，这些数据可以来源于传统行业或者互联网行业。Hadoop 生态圈中的 MapReduce 是一个大规模并行的数据计算框架，拥有很强的分布式计算能力。用户可以使用这些组件在 Hadoop 上开发和运行处理海量数据的应用程序，以满足大数据处理中的各种场景需要。

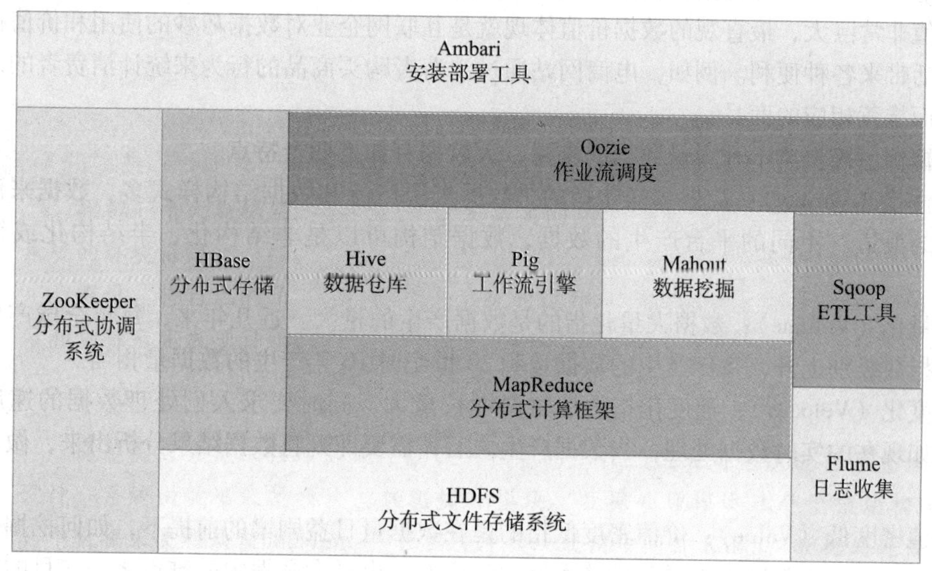


图 1 - 1 Hadoop 生态圈

低成本、高可靠、高扩展、高容错等特性让 Hadoop 成为流行的大数据分析系统，然而，Hadoop 生态圈中的 MapReduce 组件只能处理批次数据，也就是只适用于离线数据处理，在要求实时性处理的场景下毫无用处，因此，各种工具应运而生。例如，基于业务的实时性需求，有支持在线处理数据的 Storm、Cloudar Impala 及支持迭代计算的 Spark 等。同时，为了减少管理成本，提升资源的利用率，与数据分析技术一同产生的还有资源统一管理调度系统，例如，Apache Mesos、Hadoop 生态圈中的 Apache Yarn 等。

另外，基于性能、兼容性、数据类型等方面的原因，大数据分析技术不断更新，除了上面提到的各种数据处理工具之外，还有 Phoenix、Apache Accumulo、Apache Drill 等其他开源解决方案，预计未来相当长的一段时间内，主流的 Hadoop 平台将与各种新的计算模式和系统共存，并相互兼容融合，形成新一代的大数据处理系统和平台。

### 1.1.3 批数据和实时数据

大数据场景处理主要有两种数据类型，分别是批数据和流式数据。其中，批数据又称为历史数据，是在集群中积累的数据。流式数据又称为实时数据，是当前系统即时产生的数据。例如，如果把数据类比成水库，那么水库中的水就是批数据，源源不断流入水库中的水就是流式数据。

批数据和实时数据在处理延时方面还有很大差别：基于历史数据的复杂的业务查询时间一般允许在数十分钟到数小时之间，基于历史数据的交互式 SQL 查询时间一般允许数十秒到数分钟之间；基于实时数据流的数据处理延迟度要求在数百毫秒到数秒之间。无论是批数据处理还是流式数据处理，都依赖于前面提到的大数据处理技术。

### 1.1.4 数据价值挖掘

目前，整个社会对数据达成的共识可以理解成：数据像石油、煤炭资源一样宝贵，其内在的价值非常巨大，最直观的数据价值体现就是互联网企业对数据巧妙的使用和价值挖掘给人们生活带来各种便利。例如，电商网站通过消费者购买商品的行为来统计消费者的购买习惯，进而推荐相应的商品。

数据价值挖掘离不开大量数据的支撑，大数据有如下四个特点。

多样化（Variety）：数据多样化指的是数据来源广泛和数据结构样式多。数据来源可以是不同的渠道、不同的平台产生的数据，数据结构可以是非结构化、半结构化或结构化数据。

大量化（Volume）：数据大量化指的是数据产生的量大。近几年来，每年全球产生的数据量呈指数级别上升，每年产生的数据量和 20 世纪时 10 年产生的数据量相当。

速度化（Velocity）：速度化指的是数据增长量大，同时要求人们处理数据的速度也要快，比如现在的实时数据处理，当数据产生之后，需要即刻将数据结果分析出来，做到实时反馈。

价值密度低（Value）：价值密度低指的是在数据量日益剧增的前提下，如何挖掘出有用的数据。这一点正是以上数据分析技术产生的原因，也是大数据中的难点之一，目的就是要在海量的、复杂的数据中做深度分析，从中挖掘出有用的信息。

显然，无论数据量多大、什么结构、数据源头是什么，如何从海量数据中挖掘出能给使用者带来价值的数据信息，这点至关重要。大数据的核心价值在实践中主要有如下三个方面。

**1. 数据辅助决策**

通过对数据的分析，对企业提供基础的数据统计报表分析查询服务。分析师可以通过分析统计报表来指导产品生产和运营；产品经理可以通过统计数据来完善产品的功能，提高用户的体验；运营人员可以通过数据分析来发现运营问题并确定运营策略方向；管理层可以通过数据掌握公司运行情况，从而进行一些战略决策。

**2. 数据驱动业务**

通过数据分析，数据挖掘模型实现企业产品和运营的智能化，从而极大地提高企业的整体效能产出。最常见的领域有基于个性化的推荐服务和精准广告营销服务、基于模型算法的风控反欺诈服务等。

**3. 数据对外变现**

通过对数据进行精心的包装，对外提供数据服务，从而获取现金收入。市面上比较常见的有各个大数据公司利用自己掌握的大数据技术，提供风控查询验证，提供导客、导流、精准营销服务等。

## 任务 1.2　实时数据处理

### 1.2.1　实时数据的价值

上一小节中讲述了挖掘数据价值的意义，在实时数据价值中也有同样的体现。在互联网和物联网飞速发展的今天，大到天上的卫星、奔驰在路上的汽车、各种交通测速仪，小到人们手中的智能移动设备、环保监测站点、管线上某个阀门处的芯片等，无处不在的传感器时时刻刻都在产生着数据。如果可以将以上实时场景下的数据进行统计分析，则数据产生的价值远远大于数据本身。

实时数据对于大数据公司来说，价值更为重要，如实时数据帮助营销人员了解其业务和客户行为的当前影响，无论是网站流量、社会知名度还是实时广告活动。实时数据的使用弥补了在线和离线购买体验之间的差距。有实体存在的电子商务商店正在迅速采取措施为顾客实现无缝的店内体验。利用蓝牙和 GPS 的连接性，一些实体店零售商正在积极地收集关于顾客在店内行为的数据，以便向他们提供关于商店商品的个性化报价和交易。实时预测分析对于在线平台上的电子商务业务的成功越来越重要。通过挖掘实时分析的潜力，企业可以制定高度有针对性的客户参与策略，提供增强的产品推荐，甚至通过积极鼓励潜在客户完成交易来减少将产品从购物车删除的发生率。分析实时数据还可以帮助企业做出更具战略性的营销决策，并且可以改变整体业务性能。

### 1.2.2　实时数据处理技术

针对实时数据处理，目前主流的实时数据处理框架有 Storm、Spark、Samza、Flink。下面简单介绍这些技术的特点。

**1. Storm**

Storm 最开始是由 Nathan Marz 和他的团队于 2010 年在数据分析公司 BackType 开发的，后来 BackType 公司被 Twitter 收购，接着 Twitter 开源 Storm 并在 2014 年成为 Apache 顶级项目。毋庸置疑，Storm 成为大规模流数据处理的先锋，并逐渐成为工业标准。Storm 是原生的流处理系统，提供 low - level 的 API。Storm 使用 Thrift 来定义 topology 和支持多语言协议，使得人们可以使用大部分编程语言开发，Scala 自然包括在内。Trident 是对 Storm 的一个更高层次的抽象，Trident 最大的特点是以 batch 的形式进行流处理。Trident 简化 topology 构建过程，增加了窗口操作、聚合操作或者状态管理等高级操作，这些在 Storm 中并不支持。对应于 Storm 的 At most once 流传输机制，Trident 提供了 Exactly once 传输机制。Trident 支持 Java、Clojure 和 Scala 语言开发。

**2. Spark**

当前流式处理框架中，Spark 是非常受欢迎的批处理框架，包含 Spark SQL、MLlib 和

Spark Streaming。Spark 的运行是建立在批处理之上的，因此，后续加入的 Spark Streaming 也依赖于批处理，实现了微批处理，接收器把输入数据流分成短小批处理，并以类似 Spark 作业的方式处理微批处理。可以通过控制 SparkStreaming 中微批处理的时间来控制数据接收速度，同时，也可以通过外部参数控制。SparkStreaming 相对于 Storm 来说，吞吐量大，同时，可以在流批次中使用 Spark 的各种 API 扩展。Spark Streaming 提供高级声明式 API（支持 Scala、Java 和 Python 语言开发）。

**3. Samza**

最开始是专为 LinkedIn 公司开发的流处理解决方案，并和 LinkedIn 的 Kafka 一起贡献给社区，现已成为基础设施的关键部分。Samza 的构建严重依赖于基于 log 的 Kafka，两者紧密耦合。Samza 提供组合式 API，当然，也支持 Scala 语言开发。

**4. Flink**

Flink 是个相当早的项目，开始于 2008 年。Flink 是原生的流处理系统，提供 high level 的 API。Flink 也提供 API 来像 Spark 一样进行批处理，但两者处理的基础是完全不同的。Flink 把批处理当作流处理中的一种特殊情况。在 Flink 中，所有的数据都看作流，是一种很好的抽象，因为这更接近于现实世界。

### 1.2.3 Spark 实时数据处理

Spark 提供强大的内存计算引擎，几乎涵盖了所有典型的大数据计算模式，包括迭代计算、批处理计算、内存计算、流式计算（Spark Streaming）、数据查询分析计算（Shark）及图计算（GraphX）。Spark 使用 Scala 作为应用框架，采用基于内存的分布式数据集，优化了迭代式的工作负载及交互式查询。与 Hadoop 不同的是，Spark 和 Scala 紧密集成，Scala 像管理本地 collective 对象那样管理分布式数据集。Spark 支持分布式数据集上的迭代式任务，实际上也可以基于 Hadoop 生态圈中的 Yarn 资源调度框架运行。

Spark 中实时数据处理技术具体指的是 SparkStreaming。SparkStreaming 是 Spark 核心的扩展，支持实时数据处理，提供了一种抽象的连续数据流，即 Discretized Stream，简称为 DStream。Spark Streaming 在内部的处理机制是，接收实时流的数据，并根据一定的时间间隔拆分成一批一批的数据，然后通过 Spark Engine 处理这些批数据，最终得到处理后的一批一批结果数据。与其他流式处理框架相比，SparkStreaming 的优点是吞吐量大，可以人为控制接收数据时间来控制数据的流量，同时，提供高可靠的数据保证，保证数据不丢失及精准消费一次数据。

# 项目二
# Spark 集群的构建与安装

【项目描述】

模拟一个大数据实时处理项目开发需要的环境，正式开启大数据实时分析系统的环境搭建。

假设，将自己的角色定义为××系统的运维人员，我们需要进行不断的学习和训练来完成项目组分配给我们的工作。

【项目分析】

通过需求分析工程师、软件设计工程师的工作，已经对系统进行了整体的设计，系统由数据采集模块、数据缓冲模块、数据处理模块三个子模块构成，其中数据采集模块使用 Flume，数据缓冲模块使用中间件 Kafka，数据处理模块使用 SparkStreaming。系统结构如图 2－1 所示。

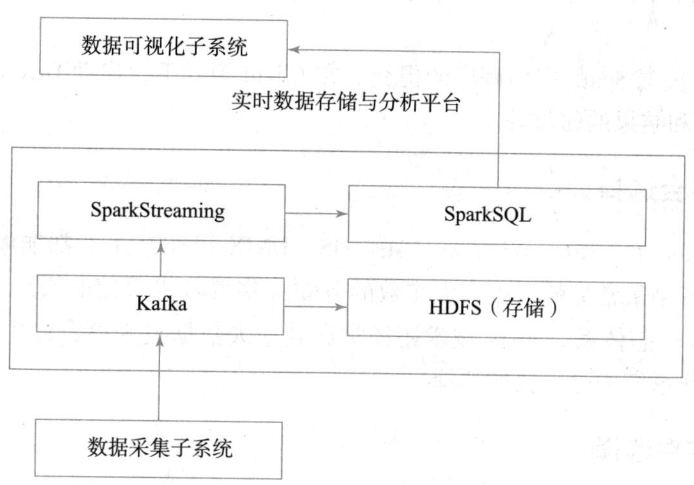


图 2－1　系统结构

作为系统运维组成员，项目经理分配的工作有如下几项：

①搭建 Spark 集群系统，为数据处理提供支撑平台（项目二内容）。

②Spark 基于 Yarn 的运行配置，为 spark on yarn 提供平台支撑（项目三内容）。

③搭建 Spark 历史日志系统，为查看日志提供平台支撑（项目四内容）。

④搭建 Kafka 消息系统，为××系统的数据缓存提供支撑平台（项目五、项目六内容）。

××系统生产环境用于支持项目的上线运行，应保证其高可靠性，并提供自动化运行能力。未来要使××系统能正常运行，需要准备 Spark 集群、Spark 日志管理系统、Kafka 分布式消息队列等环境，这里默认集群中已经安装好了部分大数据组件，例如 Hadoop、ZooKeeper、Hive、Flume 等。

根据项目经理对工作的安排，从第一项工作开始，搭建 Spark 集群系统，为数据处理提供支撑平台。

## 任务 2.1　Spark 集群搭建

完整的 Spark 集群应该包含以下三部分。

- Master：Spark 资源管理的主节点，负责管理集群的总资源、管理 Worker 节点，以及为提交在集群中的应用程序分配资源。
- Worker：Spark 资源管理的从节点，在 Cluster 模式下负责启动 Driver。负责启动 Executor 运行 task。此外，Worker 节点还负责数据的存储、数据的持久化、数据处理等。
- 客户端：负责 Spark 应用程序的提交。在 Client 模式下，启动 Driver 程序，负责监控 task 执行和结果的回收等。

### 2.1.1　平台选择

理论上，Spark 在 Linux、Windows、Mac OS、UNIX 操作系统上都能够运行。其中，基于 Linux 操作系统是最常见的，因为大多数的分布式集群技术采用的系统都是 Linux 操作系统，例如 Hadoop 生态体系。Spark 技术还经常和其他大数据技术整合使用，所以，将 Spark 安装在 Linux 操作系统中是不错的选择。

### 2.1.2　软件选择

安装 Spark 需要在各个节点上安装 JDK，因为 Spark 底层源码是用 Scala 语言和 Java 语言编写的，这两种语言都是基于 JVM 运行的。无论在哪个操作系统上运行，都需要安装 Java。

除了 Java 以外，还需要在集群内的全部主机上安装 SSH 并且保证 SSHD 进程一直运行，配置主机之间的免密登录能够更好地保证 Spark 在运行过程中不需要大量的人工介入。

### 2.1.3 Spark 集群构建流程

通常构建一个 Spark 分布式集群需要完成以下工作步骤。

**1. 规划 Spark 集群**

在安装之前，需要根据硬件条件及项目需求对集群进行整体的规划，通常包括主机规划、软件规划、网络拓扑结构规划及集群规划。

**2. 运行平台构建**

需要按照集群规划方案安装主机操作系统、修改主机名称、配置主机 IP 地址等工作。本项目假设读者能够独立安装主机操作系统。

**3. 安装配置支持软件**

需要按照集群规划方案配置集群间的免密登录，为每台主机安装 JDK。为了确保 SSHD 能够开机运行，需要额外做一些确认工作。

**4. 安装配置 Spark**

需要按照集群规划方案为每台主机安装 Spark，并将 Spark 配置为与集群规划方案一致的角色。

**5. 启动 Spark 并验证**

至此，Spark 已经按照集群规划完成了全部的安装部署。为了确保安装配置的 Spark 能够正常运行，需要启动 Spark 的各个进程，查看各节点的状态，访问 Web UI，以确定其运行正常。最后，需要运行一个 Spark 内置的 SparkPi 应用程序来验证 Spark 的功能能够正常运行。

## 任务 2.2 Spark 集群规划

本任务将学习如何安装、配置和管理 Spark 集群。首先对即将部署的 Spark 集群进行总体的规划。

在规划过程中，使用到了 5 台服务器，这里默认 5 台服务器已经修改好了主机名称，主机名称分别为 mynode1、mynode2、mynode3、mynode4、mynode5。其中 mynode1 ~ mynode4 集群节点和 Spark 集群搭建相关，mynode5 将与项目三中 Spark 基于 Yarn 运行时 Hadoop 集群相关。

### 2.2.1 Spark 集群节点划分

Spark 集群节点划分见表 2-1。

表 2-1 Spark 集群节点划分

| 主机名称 | IP 地址 | Spark 角色名称 | CPU 核心数 | 内存/GB | 硬盘容量/GB |
|---|---|---|---|---|---|
| mynode1 | 192.168.179.13 | Master | 8 核 64 位（不少于 4 核） | 20 | 500 以上 |

续表

| 主机名称 | IP 地址 | Spark 角色名称 | CPU 核心数 | 内存/GB | 硬盘容量/GB |
|---|---|---|---|---|---|
| mynode2 | 192. 168. 179. 14 | Worker1 | 8 核 64 位（不少于双核） | 20 | 500 以上 |
| mynode3 | 192. 168. 179. 15 | Worker2 | 8 核 64 位（不少于双核） | 20 | 500 以上 |
| mynode4 | 192. 168. 179. 16 | 客户端 | 8 核 64 位（不少于双核） | 20 | 500 以上 |
| mynode5 | 192. 168. 179. 17 | — | 8 核 64 位（不少于双核） | 20 | 500 以上 |

### 2.2.2 软件要求

软件要求见表 2－2。

表 2－2 软件要求

| 软件 | 版本号 | 位数/位 | 版本说明 |
|---|---|---|---|
| 操作系统 | CentOS 6. 5 | 64 | |
| JDK | 1. 8. X | 64 | |
| Spark | 2. 3. 1 | — | 稳定版本 |

除非特别说明，本书中所有的系统操作及命令均运行在 CentOS 6. 5 操作系统上，其他 Linux 操作系统的命令与 CentOS 操作系统可能略有不同，请各位读者自行查阅相关资料。

### 2.2.3 网络拓扑结构规划

网络拓扑结构规划如图 2－2 所示。

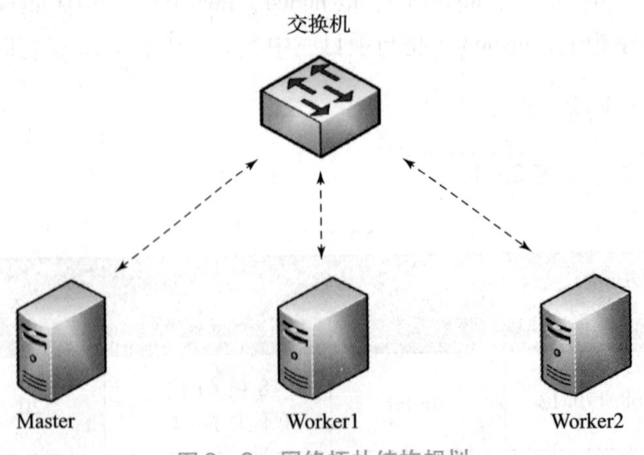


图 2－2 网络拓扑结构规划

## 任务 2.3　Spark 安装准备

### 2.3.1　JDK 安装

登录官网下载 JDK8，进行安装配置。这里下载的是 tar. gz 格式 JDK8。直接将下载好的压缩包 jdk-8u181-linux-x64. tar. gz 上传到 Master 中，进行解压配置即可。网站地址为：https://www.oracle.com/technetwork/java/javase/downloads/jdk8-downloads-2133151.html。

安装 JDK 的步骤如下：

将下载好的 JDK 上传到 192.168.179.13 Master 节点的/software/路径下，解压：

```
tar -zxvf ./jdk-8u181-linux-x64.tar.gz -C /software/
```

验证当前节点没有安装 JDK：

```
[root@mynode1 ~]# java -version
-bash: java: command not found
```

在 Master 节点上配置/etc/profile 文件，配置 JDK 环境变量：

```
[root@mynode1 software]# vim /etc/profile
#在文件最后追加如下内容：
export JAVA_HOME=/software/jdk1.8.0_181
export PATH=$PATH:$JAVA_HOME/bin
```

执行如下命令，使环境变量生效，检查 JDK 安装是否成功：

```
[root@mynode1 software]# source /etc/profile
[root@mynode1 software]# java -version
java version "1.8.0_181"
Java(TM) SE Runtime Environment (build 1.8.0_181-b13)
Java HotSpot(TM) 64-Bit Server VM (build 25.181-b13, mixed mode)
```

将解压的 JDK 和配置好的/etc/profile 文件发送到两个 Worker 节点和客户端，当将文件发送到其他节点时，需要输入对应节点的密码：

```
#将解压好的 JDK 发送到 Worker 节点和客户端
[root@mynode1 software]# scp -r ./jdk1.8.0_181/192.168.179.
14:'pwd'
[root@mynode1 software]# scp -r ./jdk1.8.0_181/192.168.179.
15:'pwd'
[root@mynode1 software]# scp -r ./jdk1.8.0_181/192.168.179.
16:'pwd'

```

```
#将配置好的profile发送到Worker节点和客户端
[root@mynode1 software]# scp /etc/profile 192.168.179.14:/etc/
[root@mynode1 software]# scp /etc/profile 192.168.179.15:/etc/
[root@mynode1 software]# scp /etc/profile 192.168.179.16:/etc/
```

在两台Worker节点和客户端节点上分别执行如下命令，使profile文件生效，并检查JDK是否安装成功：

```
#在Worker1上更新profile文件,检查JDK是否配置成功
[root@mynode2 software]# source /etc/profile
[root@mynode2 software]# java -version
java version "1.8.0_181"
Java(TM) SE Runtime Environment (build 1.8.0_181-b13)
Java HotSpot(TM) 64-Bit Server VM (build 25.181-b13, mixed mode)

#在Worker1上更新profile文件,检查JDK是否配置成功
[root@mynode3 ~]# java -version
java version "1.8.0_181"
Java(TM) SE Runtime Environment (build 1.8.0_181-b13)
Java HotSpot(TM) 64-Bit Server VM (build 25.181-b13, mixed mode)

#在客户端上更新profile文件,检查JDK是否配置成功
[root@mynode4 ~]# clear
[root@mynode4 ~]# java -version
java version "1.8.0_181"
Java(TM) SE Runtime Environment (build 1.8.0_181-b13)
Java HotSpot(TM) 64-Bit Server VM (build 25.181-b13, mixed mode)
```

各个节点显示以上信息，说明JDK配置成功。至此，JDK安装完成。

### 2.3.2 节点间的配置

在节点之间配置免密的目的是节点之间在集群启动或者任务运行过程中会有大量通信，配置了免密登录后，可以避免后期节点之间有密码访问需要人为输入密码的麻烦，同时，当节点之间有文件或者数据传输时，需要人为输入节点密码，非常不方便，配置了免密，极大地削弱了人工干预集群的情况，这是有必要的。

各个节点之间免密配置操作的步骤如下。

在Master节点、两个Worker节点、客户端上分别输入如下命令，生成密钥对：

```
#在 Master 节点上执行结果
[root@mynode1 ~]# ssh-keygen -t rsa -P ''
Generating public/private rsa key pair.
Enter file in which to save the key (/root/.ssh/id_rsa):
Your identification has been saved in /root/.ssh/id_rsa.
Your public key has been saved in /root/.ssh/id_rsa.pub.
The key fingerprint is:
ce:57:f4:69:2d:58:12:e7:7a:a7:b8:27:67:e6:d1:6e root@mynode1
The key's randomart image is:
+--[ RSA 2048]----+
|. .              |
| +               |
|o o              |
|. * o            |
|S + * o          |
|o . + =          |
|o . . o .        |
| . . . = oE      |
| .B...           |
+-----------------+

#在 Worker1 节点上执行结果
[root@mynode2 ~]# ssh-keygen -t rsa -P ''
Generating public/private rsa key pair.
Enter file in which to save the key (/root/.ssh/id_rsa):
Your identification has been saved in /root/.ssh/id_rsa.
Your public key has been saved in /root/.ssh/id_rsa.pub.
The key fingerprint is:
69:77:e8:58:83:5a:02:8d:f6:31:61:83:99:62:80:c2 root@mynode2
The key's randomart image is:
+--[ RSA 2048]----+
|+. + +           |
|oEo + + o        |
|.. . + +         |
| . o oo .        |
|o S = .          |
| = = o           |
```

```
|...              |
|                 |
|                 |
+-----------------+

#在Worker2节点上执行结果
[root@mynode3 ~]# ssh-keygen -t rsa -P ''
Generating public/private rsa key pair.
Enter file in which to save the key (/root/.ssh/id_rsa):
Your identification has been saved in /root/.ssh/id_rsa.
Your public key has been saved in /root/.ssh/id_rsa.pub.
The key fingerprint is:
b4:00:ae:b0:fd:db:68:ea:1d:5c:7e:a9:40:02:3b:ac root@mynode3
The key's randomart image is:
+--[ RSA 2048]----+
|.                |
|..               |
|o...             |
|.* .o.           |
|+.+ ..S          |
|.. = o.          |
|E = .o           |
|..* o            |
| .oo + o         |
+-----------------+

#在客户端节点上执行结果
[root@mynode4 ~]# ssh-keygen -t rsa -P ''
Generating public/private rsa key pair.
Enter file in which to save the key (/root/.ssh/id_rsa):
Your identification has been saved in /root/.ssh/id_rsa.
Your public key has been saved in /root/.ssh/id_rsa.pub.
The key fingerprint is:
ee:92:be:3e:26:3f:17:8b:32:fc:70:0b:91:b3:95:87 root@mynode4
The key's randomart image is:
+--[ RSA 2048]----+
|                 |
```

```
|                 |
|                 |
|.o               |
|+ E S            |
|= o.             |
|.+ .o.o          |
|= +Boo           |
|XB*.             |
+-----------------+
```

在每个节点上执行如下命令，将公钥文件写入授权文件中，并赋予权限：

```
#在 Master 节点上执行如下命令
[root@mynode1 ~]# cat ~/.ssh/id_rsa.pub >> ~/.ssh/authorized_keys
[root@mynode1 ~]# chmod 600 ~/.ssh/authorized keys

#在 Worker1 节点上执行如下命令
[root@mynode2 ~]# cat ~/.ssh/id_rsa.pub >> ~/.ssh/authorized_keys
[root@mynode2 ~]# chmod 600 ~/.ssh/authorized_keys

#在 Worker2 节点上执行如下命令
[root@mynode3 ~]# cat ~/.ssh/id_rsa.pub >> ~/.ssh/authorized_keys
[root@mynode3 ~]# chmod 600 ~/.ssh/authorized_keys

#在客户端节点上执行如下命令
[root@mynode4 ~]# cat ~/.ssh/id_rsa.pub >> ~/.ssh/authorized_keys
[root@mynode4 ~]# chmod 600 ~/.ssh/authorized_keys
```

节点到节点之间的无密码访问，原理是登录哪台机器，就把当前机器的公钥 copy 追加写入所登录机器的授权文件中即可。

下面配置 Master 节点无密码登录的其他节点，执行命令如下：

```
#在 Master 节点上执行如下命令
[root@mynode1 ~]# scp ~/.ssh/id_rsa.pub root@mynode2:~
root@mynode2's password: #输入对应节点密码
id_rsa.pub 100% 394 0.4KB/s 00:00
```

```
[root@mynode1 ~]# scp ~/.ssh/id_rsa.pub root@mynode3:~
root@mynode3's password: #输入对应节点密码
id_rsa.pub 100% 394 0.4KB/s 00:00
[root@mynode1 ~]# scp ~/.ssh/id_rsa.pub root@mynode4:~
root@mynode4's password: #输入对应节点密码
id_rsa.pub

#在 Worker1 节点[mynode2]上执行如下命令
[root@mynode2 ~]# cat ~/id_rsa.pub >> ~/.ssh/authorized_keys #追加公钥到认证文件
[root@mynode2 ~]# chmod 600 ~/.ssh/authorized_keys #修改认证文件权限
[root@mynode2 ~]# rm -rf ~/id_rsa.pub #删除当前节点上的公钥

#在 Worker2 节点[mynode3]上执行如下命令
[root@mynode3 ~]# cat ~/id_rsa.pub >> ~/.ssh/authorized_keys
[root@mynode3 ~]# chmod 600 ~/.ssh/authorized_keys
[root@mynode3 ~]# rm -rf ~/id_rsa.pub

#在客户端节点[mynode4]上执行如下命令
[root@mynode4 ~]# cat ~/id_rsa.pub >> ~/.ssh/authorized_keys
[root@mynode4 ~]# chmod 600 ~/.ssh/authorized_keys
[root@mynode4 ~]# rm -rf ~/id_rsa.pub

#验证 Master 节点到其他节点免密登录是否配置成功
[root@mynode1 ~]# ssh mynode2
Last login: Fri Apr 12 08:23:49 2019 from 192.168.179.1
[root@mynode2 ~]# exit
logout
Connection to mynode2 closed.
[root@mynode1 ~]# ssh mynode3
Last login: Thu Apr 11 23:05:25 2019 from 192.168.179.1
[root@mynode3 ~]# exit
logout
Connection to mynode3 closed.
[root@mynode1 ~]# ssh mynode4
Last login: Thu Apr 11 23:05:33 2019 from 192.168.179.1
```

```
[root@mynode4 ~]# exit
logout
Connection to mynode4 closed.
```

至此，Master 节点登录其他节点已经是免密的。

配置 Worker1 节点登录到其他节点免密：

```
#在 Worker1 节点上执行如下命令
[root@mynode2 ~]# scp ~/.ssh/id_rsa.pub root@mynode1:~
root@mynode1's password: #输入对应节点的密码
id_rsa.pub 100% 394 0.4KB/s 00:00
[root@mynode2 ~]# scp ~/.ssh/id_rsa.pub root@mynode3:~
root@mynode3's password: #输入对应节点的密码
id_rsa.pub 100% 394 0.4KB/s 00:00
[root@mynode2 ~]# scp ~/.ssh/id_rsa.pub root@mynode4:~
root@mynode4's password: #输入对应节点的密码
id_rsa.pub
#在 Master 节点上执行如下命令
[root@mynode1 ~]# cat ~/id_rsa.pub >> ~/.ssh/authorized_keys #追加公钥到认证文件
[root@mynode1 ~]# chmod 600 ~/.ssh/authorized_keys #修改认证文件权限
[root@mynode1 ~]# rm -rf ~/id_rsa.pub #删除当前节点上的公钥
#在 Worker2 节点上执行如下命令
[root@mynode3 ~]# cat ~/id_rsa.pub >> ~/.ssh/authorized_keys #追加公钥到认证文件
[root@mynode3 ~]# chmod 600 ~/.ssh/authorized_keys #修改认证文件权限
[root@mynode3 ~]# rm -rf ~/id_rsa.pub #删除当前节点上的公钥
#在客户端节点上执行如下命令
[root@mynode4 ~]# cat ~/id_rsa.pub >> ~/.ssh/authorized_keys #追加公钥到认证文件
[root@mynode4 ~]# chmod 600 ~/.ssh/authorized_keys #修改认证文件权限
[root@mynode4 ~]# rm -rf ~/id_rsa.pub #删除当前节点上的公钥
```

```

#验证 Worker1 节点到其他节点免密登录是否配置成功
[root@mynode2 ~]# ssh mynode1
Last login: Fri Apr 12 08:27:13 2019 from 192.168.179.14
[root@mynode1 ~]# exit
logout
Connection to mynode1 closed.
[root@mynode2 ~]# ssh mynode3
Last login: Fri Apr 12 08:24:17 2019 from 192.168.179.13
[root@mynode3 ~]# exit
logout
Connection to mynode3 closed.
[root@mynode2 ~]# ssh mynode4
Last login: Fri Apr 12 08:24:23 2019 from 192.168.179.13
[root@mynode4 ~]# exit
logout
Connection to mynode4 closed.
```

至此，Worker1 节点登录其他节点已经是免密的。

配置 Worker2 节点登录到其他节点免密：

```
#在 Worker2 节点上执行如下命令
[root@mynode3 ~]# scp ~/.ssh/id_rsa.pub root@mynode1:~
root@mynode1's password:
id_rsa.pub 100% 394 0.4KB/s 00:00
[root@mynode3 ~]# scp ~/.ssh/id_rsa.pub root@mynode2:~
root@mynode2's password:
id_rsa.pub 100% 394 0.4KB/s 00:00
[root@mynode3 ~]# scp ~/.ssh/id_rsa.pub root@mynode4:~
root@mynode4's password:
id_rsa.pub

#在 Master 节点上执行如下命令
[root@mynode1 ~]# cat ~/id_rsa.pub >> ~/.ssh/authorized_keys
#追加公钥到认证文件
[root@mynode1 ~]# chmod 600 ~/.ssh/authorized_keys
#修改认证文件权限
[root@mynode1 ~]# rm -rf ~/id_rsa.pub #删除当前节点上的公钥

```

```
#在Worker1 节点上执行如下命令
[root@mynode2 ~]# cat ~/id_rsa.pub >> ~/.ssh/authorized_keys #追加公钥到认证文件
[root@mynode2 ~]# chmod 600 ~/.ssh/authorized_keys #修改认证文件权限
[root@mynode2 ~]# rm -rf ~/id_rsa.pub #删除当前节点上的公钥

#在客户端节点上执行如下命令
[root@mynode4 ~]# cat ~/id_rsa.pub >> ~/.ssh/authorized_keys #追加公钥到认证文件
[root@mynode4 ~]# chmod 600 ~/.ssh/authorized_keys #修改认证文件权限
[root@mynode4 ~]# rm -rf ~/id_rsa.pub #删除当前节点上的公钥

#验证Worker2 节点到其他节点免密登录是否配置成功
[root@mynode3 ~]# ssh mynode1
Last login: Fri Apr 12 08:27:29 2019 from 192.168.179.14
[root@mynode1 ~]# exit
logout
Connection to mynode1 closed.
[root@mynode3 ~]# ssh mynode2
Last login: Fri Apr 12 08:27:20 2019 from 192.168.179.14
[root@mynode2 ~]# exit
logout
Connection to mynode2 closed.
[root@mynode3 ~]# ssh mynode4
Last login: Fri Apr 12 08:27:42 2019 from 192.168.179.14
[root@mynode4 ~]# exit
logout
Connection to mynode4 closed.
```

至此，Worker2 节点登录其他节点已经是免密的。

配置客户端登录到其他节点免密：

```
#在客户端节点上执行如下命令
[root@mynode4 ~]# scp ~/.ssh/id_rsa.pub root@mynode1:~
root@mynode1's password:
id_rsa.pub 100% 394 0.4KB/s 00:00
[root@mynode4 ~]# scp ~/.ssh/id_rsa.pub root@mynode2:~
```

```
root@mynode2's password:
id_rsa.pub 100% 394 0.4KB/s 00:00
[root@mynode4 ~]# scp ~/.ssh/id_rsa.pub root@mynode3:~
root@mynode3's password:
id_rsa.pub
#在 Master 节点上执行如下命令
[root@mynode1 ~]# cat ~/id_rsa.pub >> ~/.ssh/authorized_keys #追加公钥到认证文件
[root@mynode1 ~]# chmod 600 ~/.ssh/authorized_keys #修改认证文件权限
[root@mynode1 ~]# rm -rf ~/id_rsa.pub #删除当前节点上的公钥
#在 Worker1 节点上执行如下命令
[root@mynode2 ~]# cat ~/id_rsa.pub >> ~/.ssh/authorized_keys #追加公钥到认证文件
[root@mynode2 ~]# chmod 600 ~/.ssh/authorized_keys #修改认证文件权限
[root@mynode2 ~]# rm -rf ~/id_rsa.pub #删除当前节点上的公钥
#在 Worker2 节点上执行如下命令
[root@mynode3 ~]# cat ~/id_rsa.pub >> ~/.ssh/authorized_keys #追加公钥到认证文件
[root@mynode3 ~]# chmod 600 ~/.ssh/authorized_keys #修改认证文件权限
[root@mynode3 ~]# rm -rf ~/id_rsa.pub #删除当前节点上的公钥
#验证客户端节点到其他节点免密登录是否配置成功
[root@mynode4 ~]# ssh mynode1
Last login: Fri Apr 12 08:31:58 2019 from 192.168.179.15
[root@mynode1 ~]# exit
logout
Connection to mynode1 closed.
[root@mynode4 ~]# ssh mynode2
Last login: Fri Apr 12 08:32:02 2019 from 192.168.179.15
[root@mynode2 ~]# exit
logout
```

```
Connection to mynode2 closed.
[root@mynode4 ~]# ssh mynode3
Last login: Fri Apr 12 08:31:52 2019 from 192.168.179.1
[root@mynode3 ~]# exit
logout
Connection to mynode3 closed.
```

至此，客户端节点登录其他节点已经是免密的。

### 2.3.3　Spark 安装包下载

可以登录 Spark 的官网 http://spark.apache.org/downloads.html 中下载最新版本的 Spark 安装包，也可以在 https://archive.apache.org/dist/spark/下载 Spark 的历史版本。

后期会将 Spark 与 Hadoop 体系的技术进行整合，这里选择的 Spark 版本是基于Hadoop 2.6 的 spark – 2.3.1 – bin – hadoop2.6.tgz，如图 2 – 3 所示。单击之后即可下载，将下载好的压缩包保存起来。

安全 | https://archive.apache.org/dist/spark/spark-2.3.1/

**Index of /dist/spark/spark-2.3.1**

| Name | Last modified | Size | Description |
|---|---|---|---|
| Parent Directory | | - | |
| SparkR_2.3.1.tar.gz | 2018-06-01 20:59 | 303K | |
| SparkR_2.3.1.tar.gz.asc | 2018-06-01 20:59 | 819 | |
| SparkR_2.3.1.tar.gz.sha512 | 2018-06-01 20:59 | 207 | |
| pyspark-2.3.1.tar.gz | 2018-06-01 20:59 | 202M | |
| pyspark-2.3.1.tar.gz.asc | 2018-06-01 20:59 | 819 | |
| pyspark-2.3.1.tar.gz.sha512 | 2018-06-01 20:59 | 210 | |
| spark-2.3.1-bin-hadoop2.6.tgz | 2018-06-01 20:59 | 214M | |
| spark-2.3.1-bin-hadoop2.6.tgz.asc | 2018-06-01 20:59 | 819 | |
| spark-2.3.1-bin-hadoop2.6.tgz.sha512 | 2018-06-01 20:59 | 268 | |
| spark-2.3.1-bin-hadoop2.7.tgz | 2018-06-01 20:59 | 215M | |
| spark-2.3.1-bin-hadoop2.7.tgz.asc | 2018-06-01 20:59 | 819 | |
| spark-2.3.1-bin-hadoop2.7.tgz.sha512 | 2018-06-01 20:59 | 268 | |
| spark-2.3.1-bin-without-hadoop.tgz | 2018-06-01 20:59 | 147M | |
| spark-2.3.1-bin-without-hadoop.tgz.asc | 2018-06-01 20:59 | 819 | |
| spark-2.3.1-bin-without-hadoop.tgz.sha512 | 2018-06-01 20:59 | 288 | |
| spark-2.3.1.tgz | 2018-06-01 20:59 | 15M | |
| spark-2.3.1.tgz.asc | 2018-06-01 20:59 | 819 | |
| spark-2.3.1.tgz.sha512 | 2018-06-01 20:59 | 195 | |

图 2 – 3　Spark 下载

## 任务 2.4　Spark 集群搭建

将下载好的 spark-2.3.1-bin-hadoop2.6.tgz 版本通过上传工具 XFTP 上传到服务器 Master 节点的/softwar 路径下。一切准备就绪，现在可以安装 Spark 了。

首先，解压上传好的 spark-2.3.1-bin-hadoop2.6.tgz 安装包：

```
[root@mynode1 software]# tar -zxvf ./spark-2.3.1-bin-hadoop2.6.tgz
```

为了未来操作更方便，将解压后的目录 spark-2.3.1-bin-hadoop2.6 名称修改为 spark-2.3.1：

```
[root@mynode1 software]# mv spark-2.3.1-bin-hadoop2.6 spark-2.3.1
```

### 2.4.1　配置 Master 节点

在目录/software/spark-2.3.1 下进行 Master 节点的配置。进入 spark-2.3.1 下的 conf 目录，编辑 spark-env.sh，这里没有 spark-env.sh，只有 spark-env.sh.template，可以先复制一份 spark-env.sh.template，修改名称为 spark-env.sh，再进行编辑：

```
[root@mynode1 conf]# cd /software/spark-2.3.1/conf/
[root@mynode1 conf]# cp spark-env.sh.template spark-env.sh
[root@mynode1 conf]# vim spark-env.sh
```

spark-env.sh 配置编辑内容如下：

```
#!/usr/bin/env bash

#
# Licensed to the Apache Software Foundation (ASF) under one or more
# contributor license agreements. See the NOTICE file distributed with
# this work for additional information regarding copyright ownership.
# The ASF licenses this file to You under the Apache License, Version 2.0
# (the "License"); you may not use this file except in compliance with
# the License. You may obtain a copy of the License at
#
# http://www.apache.org/licenses/LICENSE-2.0
```

```
#
#Unless required by applicable law or agreed to in writing, software
#distributed under the License is distributed on an "AS IS" BASIS,
# WITHOUT WARRANTIES OR CONDITIONS OF ANY KIND, either express or implied.
# See the License for the specific language governing permissions and
# limitations under the License.
#

#This file is sourced when running various Spark programs.
#Copy it as spark-env.sh and edit that to configure Spark for your site.

#Options read when launching programs locally with
# ./bin/run-example or ./bin/spark-submit
#-HADOOP_CONF_DIR, to point Spark towards Hadoop configuration files
# - SPARK_LOCAL_IP, to set the IP address Spark binds to on this node
#-SPARK_PUBLIC_DNS, to set the public dns name of the driver program

#Options read by executors and drivers running inside the cluster
#-SPARK_LOCAL_IP, to set the IP address Spark binds to on this node
#-SPARK_PUBLIC_DNS, to set the public DNS name of the driver program
#-SPARK_LOCAL_DIRS, storage directories to use on this node for shuffle and RDD data
#-MESOS_NATIVE_JAVA_LIBRARY, to point to your libmesos.so if you use Mesos

#Options read in YARN client/cluster mode
#-SPARK_CONF_DIR, Alternate conf dir.(Default: ${SPARK_HOME}/conf)
#-HADOOP_CONF_DIR, to point Spark towards Hadoop configuration files
#-YARN_CONF_DIR, to point Spark towards YARN configuration files when you use YARN
```

```
# - SPARK_EXECUTOR_CORES, Number of cores for the executors (Default: 1).
# - SPARK_EXECUTOR_MEMORY, Memory per Executor (e.g. 1000M, 2G) (Default: 1G)
# - SPARK_DRIVER_MEMORY, Memory for Driver (e.g. 1000M, 2G) (Default: 1G)
exportJAVA_HOME = /software/jdk1.8.0_181#配置 Java 环境变量
export SPARK_MASTER_HOST =mynode1#配置 Master 的主节点
export SPARK_MASTER_PORT =7077#配置集群提交任务的端口 port

# Options for the daemons used in the standalone deploy mode
# - SPARK_MASTER_HOST, to bind the master to a different IP address or hostname
# - SPARK_MASTER_PORT / SPARK_MASTER_WEBUI_PORT, to use non-default ports for the master
# - SPARK_MASTER_OPTS, to set config properties only for the master (e.g. "-Dx=y")
# - SPARK_WORKER_CORES, to set the number of cores to use on this machine
# - SPARK_WORKER_MEMORY, to set how much total memory workers have to give executors (e.g. 1000M, 2G)
# - SPARK_WORKER_PORT / SPARK_WORKER_WEBUI_PORT, to use non-default ports for the worker
# - SPARK_WORKER_DIR, to set the working directory of worker processes
# - SPARK_WORKER_OPTS, to set config properties only for the worker (e.g. "-Dx=y")
# - SPARK_DAEMON_MEMORY, to allocate to the master, worker and history server themselves (default: 1G).
# - SPARK_HISTORY_OPTS, to set config properties only for the history server (e.g. "-Dx=y")
# - SPARK_SHUFFLE_OPTS, to set config properties only for the external shuffle service (e.g. "-Dx=y")
# - SPARK_DAEMON_JAVA_OPTS, to set config properties for all daemons (e.g. "-Dx=y")
# - SPARK_DAEMON_CLASSPATH, to set the classpath for all daemons
# - SPARK_PUBLIC_DNS, to set the public dns name of the master or workers
```

```
# Generic options for the daemons used in the standalone deploy mode
# - SPARK_CONF_DIR Alternate conf dir. (Default: ${SPARK_HOME}/conf)
# - SPARK_LOG_DIR Where log files are stored. (Default: ${SPARK_HOME}/logs)
# - SPARK_PID_DIR Where the pid file is stored. (Default: /tmp)
# - SPARK_IDENT_STRING A string representing this instance of spark. (Default: $USER)
# - SPARK_NICENESS The scheduling priority for daemons. (Default: 0)
# - SPARK_NO_DAEMONIZE Run the proposed command in the foreground. It will not output a PID file.
# Options for native BLAS, like Intel MKL, OpenBLAS, and so on.
# You might get better performance to enable these options if using native BLAS (see SPARK-21305).
# - MKL_NUM_THREADS=1 Disable multi-threading of Intel MKL
# - OPENBLAS_NUM_THREADS=1 Disable multi-threading of OpenBLAS
```

上面配置了以下几个信息：

- JAVA_HOME：Spark 集群的运行需要依赖 Java 的环境变量，所以这里配置 JAVA_HOME。
- SPARK_MASTER_HOST：配置 Master 节点的主机位置。
- SPARK_MASTER_PORT：配置集群提交任务端口。

至此，Master 节点配置完成，下面开始配置 Worker 节点信息。

### 2.4.2 配置 Worker 节点

同样，编辑/software/spark-2.3.1/conf/spark-env.sh 文件，在后面追加如下配置：

```
export SPARK_WORKER_CORES=8 #配置 Worker 节点使用的核心数
export SPARK_WORKER_MEMORY=20G #配置 Worker 节点使用的内存大小
```

以上配置中配置的信息解释如下：

- SPARK_WORKER_CORES：配置 Worker 节点所能支配的核心数，这里的核心数决定这未来在 Spark Worker 节点上能并行运行的 task 个数。
- SPARK_WORKER_MEMORY：配置 Worker 节点所能使用的内存大小，内存越大越好，给足 Worker 内存是保证 Spark 运行任务顺利的前提。

除了上面的配置，还需编辑/software/spark-2.3.1/conf/slaves 文件，配置将哪些节点指定为 Worker 节点。conf 目录下没有 slaves 文件，只有 slaves. template 文件，可以复制一份 slaves. template 文件，修改名称为 slaves 并编辑：

```
[root@mynode1 conf]# cd /software/spark-2.3.1/conf/
[root@mynode1 conf]# cp slaves.template slaves
[root@mynode1 conf]# vim slaves
```

slaves 的编辑内容如下：

```
# Licensed to the Apache Software Foundation (ASF) under one or more
# contributor license agreements. See the NOTICE file distributed with
# this work for additional information regarding copyright ownership.
# The ASF licenses this file to You under the Apache License, Version 2.0
# (the "License"); you may not use this file except in compliance with
# the License. You may obtain a copy of the License at
#
# http://www.apache.org/licenses/LICENSE-2.0
#
# Unless required by applicable law or agreed to in writing, software
# distributed under the License is distributed on an "AS IS" BASIS,
# WITHOUT WARRANTIES OR CONDITIONS OF ANY KIND, either express or implied.
# See the License for the specific language governing permissions and
# limitations under the License.
#

# A Spark Worker will be started on each of the machines listed below.

mynode2
mynode3
```

slaves 文件中编辑的内容是指定“mynode2”“mynode3”节点为 Spark 集群中的 Worker 节点，未来提交到 Spark 集群中的任务就可以运行在这些 Worker 节点上。

至此，Spark 集群搭建 Worker 节点配置完成。

### 2.4.3 分发安装包

将以上配置好的 Spark-2.3.1 文件分发到各个 Worker 节点，执行如下命令：

```
scp -r /software/spark-2.3.1 root@mynode2:'pwd'
scp -r /software/spark-2.3.1 root@mynode3:'pwd'
```

在 mynode2 上检查 spark-2.3.1 安装包是否发送成功：

```
[root@mynode2 software]# cd /software/
[root@mynode2 software]# ll
drwxrwxr-x.13 root root 4096 4月 13 07:57 spark-2.3.1
```

在 mynode3 上检查 spark-2.3.1 安装包是否发送成功：

```
[root@mynode3 software]# cd /software/
[root@mynode3 software]#ll
drwxrwxr-x.13 root root 4096 4月 13 07:58 spark-2.3.1
```

经过以上步骤，Spark 集群的搭建正式完成。下面可以启动集群，在集群中提交任务，验证集群搭建是否成功。

### 2.4.4 启动集群

以上步骤执行完之后，可以启动集群，测试集群是否搭建成功。进入 Master 节点/software/spark-2.3.1/sbin 路径，执行如下命令启动集群：

```
[root@mynode1 sbin]# cd /software/spark-2.3.1/sbin/
[root@mynode1 sbin]# ./start-all.sh
starting org.apache.spark.deploy.master.Master, logging to/software/spark-2.3.1/logs/spark-root-org.apache.spark.deploy.master.Master-1-mynode1.out
mynode2: starting org.apache.spark.deploy.worker.Worker, logging to /software/spark-2.3.1/logs/spark-root-org.apache.spark.deploy.worker.Worker-1-mynode2.out
mynode3: starting org.apache.spark.deploy.worker.Worker, logging to /software/spark-2.3.1/logs/spark-root-org.apache.spark.deploy.worker.Worker-1-mynode3.out
```

## 任务 2.5 验证 Spark

### 2.5.1 验证 Spark 进程

由于 Spark 中启动的 Master、Worker 进程都是 JVM 进程，所以可以通过在相应的节点上执行 jps 命令来查看对应的进程是否存在。

进入 Master 节点，执行 jps 命令，出现 Master 进程，代表 Spark 的 Master 进程启动成功：

```
[root@mynode1 sbin]# jps
1331 Master
1412 Jps
```

进入 Worker1 节点，执行 jps 命令，出现 Worker 进程，代表 Spark 的 Worker 进程启动成功：

```
[root@mynode2 ~]# jps
1546 Jps
1483 Worker
```

进入 Worker2 节点，执行 jps 命令，出现 Worker 进程，代表 Spark 的 Worker 进程启动成功：

```
[root@mynode3 ~]# jps
1328 Worker
1391 Jps
```

经过以上各个节点验证对应的进程，发现对应的节点上启动着相应的进程，说明 Spark 集群启动成功。

### 2.5.2 验证 WEBUI 界面

Spark 集群启动之后，除了验证对应节点上的进程是否存在外，还可以验证 Spark WEBUI 是否正常。Spark WEBUI 地址为"http://Master 节点:port"，默认的 port 是 8080，这里直接在浏览器中登录 http://mynode1:8080 即可，显示如图 2－4 所示，集群启动成功。

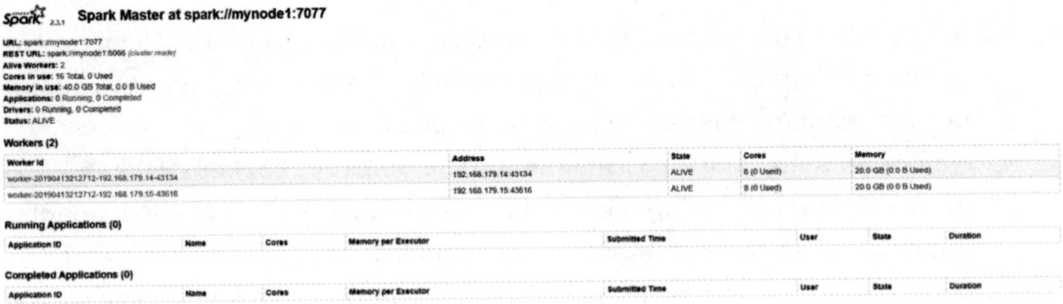

Spark 2.3.1 Spark Master at spark://mynode1:7077

URL: spark://mynode1:7077
REST URL: spark://mynode1:6066 (cluster mode)
Alive Workers: 2
Cores in use: 16 Total, 0 Used
Memory in use: 40.0 GB Total, 0.0 B Used
Applications: 0 Running, 0 Completed
Drivers: 0 Running, 0 Completed
Status: ALIVE

Workers (2)

| Worker Id | Address | State | Cores | Memory |
|---|---|---|---|---|
| worker-20190413212712-192.168.179.14-43134 | 192.168.179.14:43134 | ALIVE | 8 (0 Used) | 20.0 GB (0.0 B Used) |
| worker-20190413212712-192.168.179.15-43616 | 192.168.179.15:43616 | ALIVE | 8 (0 Used) | 20.0 GB (0.0 B Used) |

Running Applications (0)

| Application ID | Name | Cores | Memory per Executor | Submitted Time | User | State | Duration |
|---|---|---|---|---|---|---|---|

Completed Applications (0)

| Application ID | Name | Cores | Memory per Executor | Submitted Time | User | State | Duration |
|---|---|---|---|---|---|---|---|

图 2－4　Spark WEBUI 界面

### 2.5.3 验证集群功能

运行一个 Spark 源码包中的 SparkPi 案例，来验证集群运行是否正常。SparkPi 案例就是一个计算 Pi 结果的程序。SparkPi 任务的主类如下：

org.apache.spark.examples.SparkPi

以上 SparkPi 主类对应的 jar 包如下：

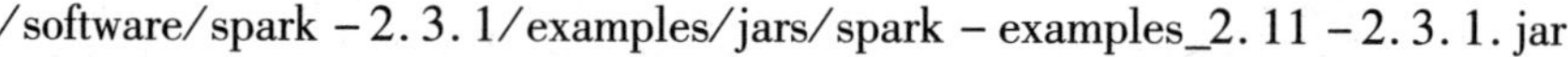
/software/spark-2.3.1/examples/jars/spark-examples_2.11-2.3.1.jar

可以选择 Master、两台 Worker 的任意节点来提交 SparkPi 任务。下面在 Master 节点中提交 SparkPi 任务。首先进入 Master 节点/software/spark-2.3.1/bin/路径，使用 Spark-submit 提交任务。操作如下：

```
[root@mynode1 sbin]# cd /software/spark-2.3.1/bin/
[root@mynode1 bin]# ./spark-submit --master spark://mynode1:7077 --class org.apache.spark.examples.SparkPi ../examples/jars/spark-examples_2.11-2.3.1.jar
```

执行过程和结果如下：

```
[root@mynode1 bin]# ./spark-submit --master spark://mynode1:7077 --class org.apache.spark.examples.SparkPi ../examples/jars/spark-examples_2.11-2.3.1.jar
2019-04-13 21:54:26 WARN NativeCodeLoader:62 - Unable to load native-hadoop library for your platform... using builtin-java classes where applicable
2019-04-13 21:54:27 INFO SparkContext:54 - Running Spark version 2.3.1
2019-04-13 21:54:27 INFO SparkContext:54 - Submitted application: Spark Pi
2019-04-13 21:54:27 INFO SecurityManager:54 - Changing view acls to: root
2019-04-13 21:54:27 INFO SecurityManager:54 - Changing modify acls to: root
2019-04-13 21:54:27 INFO SecurityManager:54 - Changing view acls groups to:
2019-04-13 21:54:27 INFO SecurityManager:54 - Changing modify acls groups to:
2019-04-13 21:54:27 INFO SecurityManager:54 - SecurityManager: authentication disabled; ui acls disabled; users with view permissions: Set(root); groups with view permissions: Set(); user
s with modify permissions: Set(root); groups with modify permissions: Set()2019-04-13 21:54:27 INFO Utils:54 - Successfully started service 'sparkDriver' on port 58497.
2019-04-13 21:54:27 INFO SparkEnv:54 - Registering MapOutputTracker
2019-04-13 21:54:27 INFO SparkEnv:54 - Registering BlockManagerMaster
```

```
13. 2019 - 04 - 13 21:54:27 INFO BlockManagerMasterEndpoint:54 -
    Using org.apache.spark.storage.DefaultTopologyMapper for get-
    ting topology information
14. 2019 - 04 - 13 21:54:27 INFO BlockManagerMasterEndpoint:54 -
    BlockManagerMasterEndpoint up
15. 2019 - 04 - 13 21:54:27 INFO DiskBlockManager:54 - Created local
    directory at /tmp/blockmgr - 0ab3f71f - 9503 - 4eef - 8a81 -
    a033483b0fff
16. 2019 - 04 - 13 21:54:27 INFO MemoryStore:54 - MemoryStore started
    with capacity 366.3 MB
17. 2019 - 04 - 13 21:54:27 INFO SparkEnv:54 - Registering OutputCom-
    mitCoordinator
18. 2019 - 04 - 13 21:54:27 INFO log:192 - Logging initialized @1790ms
19. 2019 - 04 - 13 21:54:27 INFO Server:346 - jetty - 9.3.z - SNAPSHOT
20. 2019 - 04 - 13 21:54:27 INFO Server:414 - Started @1868ms
21. 2019 - 04 - 13 21:54:27 INFO AbstractConnector:278 - Started
    ServerConnector@363d42a6{HTTP/1.1,[http/1.1]}{0.0.0.0:4040}
22. 2019 - 04 - 13 21:54:27 INFO Utils:54 - Successfully started serv-
    ice 'SparkUI' on port 4040.
23. 2019 - 04 - 13 21:54:27 INFO ContextHandler:781 - Started o.s.j.s.
    ServletContextHandler@4362d7df{/jobs,null,AVAILABLE,@Spark}
24. 2019 - 04 - 13 21:54:27 INFO ContextHandler:781 - Started o.s.j.s.
    ServletContextHandler@473b3b7a{/jobs/json,null,AVAILABLE,
    @Spark}
25. 2019 - 04 - 13 21:54:27 INFO ContextHandler:781 - Started o.s.j.s.
    ServletContextHandler@1734f68{/jobs/job,null,AVAILABLE,
    @Spark}
26. 2019 - 04 - 13 21:54:27 INFO ContextHandler:781 - Started o.s.j.s.
    ServletContextHandler@5ed190be{/jobs/job/json,null,AVAILA-
    BLE,@Spark}
27. 2019 - 04 - 13 21:54:27 INFO ContextHandler:781 - Started o.s.j.s.
    ServletContextHandler@402f80f5{/stages,null,AVAILABLE,
    @Spark}
28. 2019 - 04 - 13 21:54:27 INFO ContextHandler:781 - Started o.s.j.s.
    ServletContextHandler@5bbc9f97{/stages/json,null,AVAILABLE,
    @Spark}
29. 2019 - 04 - 13 21:54:27 INFO ContextHandler:781 - Started o.s.j.s.
```

```
   ServletContextHandler@133e019b{/stages/stage,null,AVAILA-
   BLE,@Spark}
30.2019-04-13 21:54:27 INFO ContextHandler:781 - Started o.s.j.s.
   ServletContextHandler@425357dd{/stages/stage/json,null,
   AVAILABLE,@Spark}
31.2019-04-13 21:54:27 INFO ContextHandler:781 - Started o.s.j.s.
   ServletContextHandler@2102a4d5{/stages/pool,null,AVAILABLE,
   @Spark}
32.2019-04-13 21:54:27 INFO ContextHandler:781 - Started o.s.j.s.
   ServletContextHandler@210386e0{/stages/pool/json,null,
   AVAILABLE,@Spark}
33.2019-04-13 21:54:27 INFO ContextHandler:781 - Started o.s.j.s.
   ServletContextHandler@3d4d3fe7{/storage,null,AVAILABLE,
   @Spark}
34.2019-04-13 21:54:27 INFO ContextHandler:781 - Started o.s.j.s.
   ServletContextHandler@65f87a2c{/storage/json,null,AVAILA-
   BLE,@Spark}
35.2019-04-13 21:54:27 INFO ContextHandler:781 - Started o.s.j.s.
   ServletContextHandler@51684e4a{/storage/rdd,null,AVAILABLE,
   @Spark}
36.2019-04-13 21:54:27 INFO ContextHandler:781 - Started o.s.j.s.
   ServletContextHandler@6ce1f601{/storage/rdd/json,null,
   AVAILABLE,@Spark}
37.2019-04-13 21:54:27 INFO ContextHandler:781 - Started o.s.j.s.
   ServletContextHandler@38875e7d{/environment,null,AVAILABLE,
   @Spark}
38.2019-04-13 21:54:27 INFO ContextHandler:781 - Started o.s.j.s.
   ServletContextHandler@1e886a5b{/environment/json,null,
   AVAILABLE,@Spark}
39.2019-04-13 21:54:27 INFO ContextHandler:781 - Started o.s.j.s.
   ServletContextHandler@d816dde{/executors,null,AVAILABLE,
   @Spark}
40.2019-04-13 21:54:27 INFO ContextHandler:781 - Started o.s.j.s.
   ServletContextHandler@6e33c391{/executors/json,null,AVAILA-
   BLE,@Spark}
41.2019-04-13 21:54:27 INFO ContextHandler:781 - Started o.s.j.s.
   ServletContextHandler@6c451c9c{/executors/threadDump,null,
   AVAILABLE,@Spark}
```

```
42. 2019-04-13 21:54:27 INFO ContextHandler:781 - Started o.s.j.s.
    ServletContextHandler@31c269fd{/executors/threadDump/json,
    null,AVAILABLE,@Spark}
43. 2019-04-13 21:54:27 INFO ContextHandler:781 - Started o.s.j.s.
    ServletContextHandler@372b0d86{/static,null,AVAILABLE,@
    Spark}
44. 2019-04-13 21:54:27 INFO ContextHandler:781 - Started o.s.j.s.
    ServletContextHandler@29539e36{/,null,AVAILABLE,@Spark}
45. 2019-04-13 21:54:27 INFO ContextHandler:781 - Started o.s.j.s.
    ServletContextHandler@32f61a31{/api,null,AVAILABLE,@Spark}
46. 2019-04-13 21:54:27 INFO ContextHandler:781 - Started o.s.j.s.
    ServletContextHandler@51a06cbe{/jobs/job/kill,null,AVAILA-
    BLE,@Spark}
47. 2019-04-13 21:54:27 INFO ContextHandler:781 - Started o.s.j.s.
    ServletContextHandler@3dddbe65{/stages/stage/kill,null,
    AVAILABLE,@Spark}
48. 2019-04-13 21:54:27 INFO SparkUI:54 - Bound SparkUI to 0.0.0.0,
    and started at http://mynode1:4040
49. 2019-04-13 21:54:28 INFO SparkContext:54 - Added JAR file:/
    software/spark-2.3.1/bin/../examples/jars/spark-examples_
    2.11-2.3.1.jar at spark://mynode1:58497/jars/spark-examples_
    2.11-2.3.1.jar with timestamp 1555163668031
50. 2019-04-13 21:54:28 INFO StandaloneAppClient$ClientEndpoint:
    54 - Connecting to master spark://mynode1:7077...
51. 2019-04-13 21:54:28 INFO TransportClientFactory:267 - Success-
    fully created connection to mynode1/192.168.179.13:7077 after
    64 ms (0 ms spent in bootstraps)
52. 2019-04-13 21:54:28 INFO StandaloneSchedulerBackend:54 - Con-
    nected to Spark cluster with app ID app-20190413215428-0000
53. 2019-04-13 21:54:28 INFO StandaloneAppClient$ClientEndpoint:
    54 - Executor added: app-20190413215428-0000/0 on worker-
    20190413212712-192.168.179.15-43616 (192.168.179.15:43616)
    with 8 core(s)
54. 2019-04-13 21:54:28 INFO StandaloneSchedulerBackend:54 - Gran-
    ted executor ID app-20190413215428-0000/0 on hostPort 192.
    168.179.15:43616 with 8 core(s), 1024.0 MB RAM
```

```
55. 2019-04-13 21:54:28 INFO Utils:54 - Successfully started service 'org.apache.spark.network.netty.NettyBlockTransferService' on port 49833.
56. 2019-04-13 21:54:28 INFO NettyBlockTransferService:54 - Server created on mynode1:49833
57. 2019-04-13 21:54:28 INFO StandaloneAppClient$ClientEndpoint:54 - Executor added: app-20190413215428-0000/1 on worker-20190413212712-192.168.179.14-43134 (192.168.179.14:43134) with 8 core(s)
58. 2019-04-13 21:54:28 INFO StandaloneSchedulerBackend:54 - Granted executor ID app-20190413215428-0000/1 on hostPort 192.168.179.14:43134 with 8 core(s), 1024.0 MB RAM
59. 2019-04-13 21:54:28 INFO BlockManager:54 - Using org.apache.spark.storage.RandomBlockReplicationPolicy for block replication policy
60. 2019-04-13 21:54:28 INFO StandaloneAppClient$ClientEndpoint:54 - Executor updated: app-20190413215428-0000/1 is now RUNNING
61. 2019-04-13 21:54:28 INFO StandaloneAppClient$ClientEndpoint:54 - Executor updated: app-20190413215428-0000/0 is now RUNNING
62. 2019-04-13 21:54:28 INFO BlockManagerMaster:54 - Registering BlockManagerBlockManagerId(driver, mynode1, 49833, None)
63. 2019-04-13 21:54:28 INFO BlockManagerMasterEndpoint:54 - Registering block manager mynode1:49833 with 366.3 MB RAM, BlockManagerId(driver, mynode1, 49833, None)
64. 2019-04-13 21:54:28 INFO BlockManagerMaster:54 - Registered BlockManagerBlockManagerId(driver, mynode1, 49833, None)
65. 2019-04-13 21:54:28 INFO BlockManager:54 - Initialized BlockManager: BlockManagerId(driver, mynode1, 49833, None)
66. 2019-04-13 21:54:29 INFO ContextHandler:781 - Started o.s.j.s.ServletContextHandler@6f099cef{/metrics/json,null,AVAILABLE,@Spark}
67. 2019-04-13 21:54:29 INFO StandaloneSchedulerBackend:54 - SchedulerBackend is ready for scheduling beginning after reached minRegisteredResourcesRatio: 0.0
68. 2019-04-13 21:54:29 INFO SparkContext:54 - Starting job: reduce at SparkPi.scala:38
69. 2019-04-13 21:54:29 INFO DAGScheduler:54 - Got job 0 (reduce at SparkPi.scala:38) with 2 output partitions
```

```
70. 2019-04-13 21:54:29 INFO DAGScheduler:54 - Final stage: Result-
    Stage 0 (reduce at SparkPi.scala:38)
71. 2019-04-13 21:54:29 INFO DAGScheduler:54 - Parents of final
    stage: List()
72. 2019-04-13 21:54:29 INFO DAGScheduler:54 - Missing parents:
    List()
73. 2019-04-13 21:54:29 INFO DAGScheduler:54 - Submitting Result-
    Stage 0 (MapPartitionsRDD[1] at map at SparkPi.scala:34), which
    has no missing parents
74. 2019-04-13 21:54:30 INFO MemoryStore:54 - Block broadcast_0
    stored as values in memory (estimated size 1832.0 B, free 366.3
    MB)
75. 2019-04-13 21:54:30 INFO MemoryStore:54 - Block broadcast_0_
    piece0 stored as bytes in memory (estimated size 1181.0 B, free
    366.3 MB)
76. 2019-04-13 21:54:30 INFO BlockManagerInfo:54 - Added broadcast
    _0_piece0 in memory on mynode1:49833 (size: 1181.0 B, free: 366.3
    MB)
77. 2019-04-13 21:54:30 INFO SparkContext:54 - Created broadcast 0
    from broadcast at DAGScheduler.scala:1039
78. 2019-04-13 21:54:30 INFO DAGScheduler:54 - Submitting 2 missing
    tasks from ResultStage 0 (MapPartitionsRDD[1] at map at SparkPi.
    scala:34) (first 15 tasks are for partitions Vector(0, 1))
79. 2019-04-13 21:54:30 INFO TaskSchedulerImpl:54 - Adding task
    set 0.0 with 2 tasks
80. 2019-04-13 21:54:31 INFO CoarseGrainedSchedulerBackend
    $DriverEndpoint:54 - Registered executor NettyRpcEndpointRef
    (spark-client://Executor) (192.168.179.15:54886) with ID 0
81. 2019-04-13 21:54:31 INFO TaskSetManager:54 - Starting task 0.0
    in stage 0.0 (TID 0, 192.168.179.15, executor 0, partition 0,
    PROCESS_LOCAL, 7857 bytes)
82. 2019-04-13 21:54:31 INFO TaskSetManager:54 - Starting task 1.0
    in stage 0.0 (TID 1, 192.168.179.15, executor 0, partition 1,
    PROCESS_LOCAL, 7857 bytes)
83. 2019-04-13 21:54:31 INFO CoarseGrainedSchedulerBackend
    $DriverEndpoint:54 - Registered executor NettyRpcEndpointRef
    (spark-client://Executor) (192.168.179.14:57471) with ID 1
```

```
84. 2019-04-13 21:54:31 INFO BlockManagerMasterEndpoint:54 - Registering block manager 192.168.179.15:33211 with 366.3 MB RAM, BlockManagerId(0, 192.168.179.15, 33211, None)
85. 2019-04-13 21:54:31 INFO BlockManagerMasterEndpoint:54 - Registering block manager 192.168.179.14:34462 with 366.3 MB RAM, BlockManagerId(1, 192.168.179.14, 34462, None)
86. 2019-04-13 21:54:32 INFO BlockManagerInfo:54 - Added broadcast_0_piece0 in memory on 192.168.179.15:33211 (size: 1181.0 B, free: 366.3 MB)
87. 2019-04-13 21:54:33 INFO TaskSetManager:54 - Finished task 1.0 in stage 0.0 (TID 1) in 1385 ms on 192.168.179.15 (executor 0) (1/2)
88. 2019-04-13 21:54:33 INFO TaskSetManager:54 - Finished task 0.0 in stage 0.0 (TID 0) in 1426 ms on 192.168.179.15 (executor 0) (2/2)
89. 2019-04-13 21:54:33 INFO TaskSchedulerImpl:54 - Removed TaskSet 0.0, whose tasks have all completed, from pool
90. 2019-04-13 21:54:33 INFO DAGScheduler:54 - ResultStage 0 (reduce at SparkPi.scala:38) finished in 3.137 s
91. 2019-04-13 21:54:33 INFO DAGScheduler:54 - Job 0 finished: reduce at SparkPi.scala:38, took 3.260470 s
92. Pi is roughly 3.1377356886784433 #执行结果
93. 2019-04-13 21:54:33 INFO AbstractConnector:318 - Stopped Spark@363d42a6{HTTP/1.1,[http/1.1]}{0.0.0.0:4040}
94. 2019-04-13 21:54:33 INFO SparkUI:54 - Stopped Spark web UI at http://mynode1:4040
95. 2019-04-13 21:54:33 INFO StandaloneSchedulerBackend:54 - Shutting down all executors
96. 2019-04-13 21:54:33 INFO CoarseGrainedSchedulerBackend$DriverEndpoint:54 - Asking each executor to shut down
97. 2019-04-13 21:54:33 INFO MapOutputTrackerMasterEndpoint:54 - MapOutputTrackerMasterEndpoint stopped!
98. 2019-04-13 21:54:33 INFO MemoryStore:54 - MemoryStore cleared
99. 2019-04-13 21:54:33 INFO BlockManager:54 - BlockManager stopped
100. 2019-04-13 21:54:33 INFO BlockManagerMaster:54 - BlockManagerMaster stopped
101. 2019-04-13 21:54:33 INFO OutputCommitCoordinator$OutputCommitCoordinatorEndpoint:54 - OutputCommitCoordinator stopped!
102. 2019-04-13 21:54:33 INFO SparkContext:54 - Successfully stopped SparkContext
```

```
103.2019 -04 -13 21:54:33 INFO ShutdownHookManager:54 - Shutdown
    hook called
104.2019 -04 -13 21:54:33 INFO ShutdownHookManager:54 - Deleting di-
    rectory /tmp/spark -12243bf0 - a531 - 4d1e - b81f - 56552a5c508b
105.2019 -04 -13 21:54:33 INFO ShutdownHookManager:54 - Deleting di-
    rectory /tmp/spark -9ef2c0fb - 3e9e - 4338 - a387 - 5b8ee9c499e8
```

以上是在 Spark Master 节点上执行任务产生的日志信息和结果。也可以查看对应的 Spark WEBUI，如图 2-5 所示，执行任务显示成功。

Spark 2.3.1 Spark Master at spark://mynode1:7077

**URL:** spark://mynode1:7077
**REST URL:** spark://mynode1:6066 *(cluster mode)*
**Alive Workers:** 2
**Cores in use:** 16 Total, 0 Used
**Memory in use:** 40.0 GB Total, 0.0 B Used
**Applications:** 0 Running, 1 Completed
**Drivers:** 0 Running, 0 Completed
**Status:** ALIVE

**Workers (2)**

| Worker Id | Address | State | Cores | Memory |
|---|---|---|---|---|
| worker-20190413212712-192.168.179.14-43134 | 192.168.179.14:43134 | ALIVE | 8 (0 Used) | 20.0 GB (0.0 B Used) |
| worker-20190413212712-192.168.179.15-43616 | 192.168.179.15:43616 | ALIVE | 8 (0 Used) | 20.0 GB (0.0 B Used) |

**Running Applications (0)**

| Application ID | Name | Cores | Memory per Executor | Submitted Time | User | State | Duration |
|---|---|---|---|---|---|---|---|

**Completed Applications (1)**

| Application ID | Name | Cores | Memory per Executor | Submitted Time | User | State | Duration |
|---|---|---|---|---|---|---|---|
| app-20190413215428-0000 | Spark Pi | 16 | 1024.0 MB | 2019/04/13 21:54:28 | root | FINISHED | 5 s |

图 2-5 任务执行结果

## 任务 2.6 Spark 客户端搭建 

上面提到的 SparkPi 任务在 Spark 集群中的 Master 节点和 Worker 节点都是能运行提交的，但是每次运行 Spark 任务时，当前节点都会启动一个 JVM 来提交当前任务，这个 JVM 进程会占用当前节点的资源及与 Worker 有大量的网络通信。为了不占用每一个节点的资源去提交任务，可以在一台新的节点上提交 Spark 任务。

### 2.6.1 客户端介绍

Spark 的客户端就是在提交 Spark 任务时，在单独的一台客户端上提交 Spark 任务。提交任务时，只需要指定 Spark 集群中的 Master 节点及提交任务的端口。这样每次提交任务时，任务启动的 JVM 不会占用任何 Spark 集群的资源。

### 2.6.2 客户端搭建及使用

搭建 Spark 客户端时，可以选择一台服务器节点，这里选择的是 mynode4［192.168.179.16］节点，在 2.4.1 节中已有介绍。搭建 Spark 客户端时，只需要将 Spark 的安装包上传至客户端节点解压即可。

这里将 spark -2.3.1 - bin - hadoop2.6.tgz 压缩包上传至客户端/software 下进行解压并修改名称为 spark -2.3.1：

```
[root@mynode4 software]# tar -zxvf ./spark-2.3.1-bin-hadoop2.6.tgz
[root@mynode4 software]# mv spark-2.3.1-bin-hadoop2.6 spark-2.3.1
```

在客户端上提交 SparkPi 任务：

在保证 Spark 集群启动的情况下，进入客户端节点/software/spark-2.3.1/bin/路径，提交任务：

```
[root@mynode4 software]# cd /software/spark-2.3.1/bin/
[root@mynode4 bin]# ./spark-submit --master spark://mynode1:7077 --class org.apache.spark.examples.SparkPi ../examples/jars/spark-examples_2.11-2.3.1.jar
```

执行结果如下：

```
[root@mynode4 bin]# ./spark-submit --master spark://mynode1:7077 --class org.apache.spark.examples.SparkPi ../examples/jars/spark-examples_2.11-2.3.1.jar
2019-04-13 22:55:54 WARN NativeCodeLoader:62 - Unable to load native-hadoop library for your platform... using builtin-java classes where applicable
2019-04-13 22:55:54 INFO SparkContext:54 - Running Spark version 2.3.1
2019-04-13 22:55:54 INFO SparkContext:54 - Submitted application: Spark Pi
2019-04-13 22:55:54 INFO SecurityManager:54 - Changing view acls to: root
2019-04-13 22:55:54 INFO SecurityManager:54 - Changing modify acls to: root
2019-04-13 22:55:54 INFO SecurityManager:54 - Changing view acls groups to:
2019-04-13 22:55:54 INFO SecurityManager:54 - Changing modify acls groups to:
2019-04-13 22:55:54 INFO SecurityManager:54 - SecurityManager: authentication disabled; uiacls disabled; users with view permissions: Set(root); groups with view permissions: Set(); user s with modify permissions: Set(root); groups with modify permissions: Set()
2019-04-13 22:55:55 INFO Utils:54 - Successfully started service 'sparkDriver' on port 50000.
```

```
11. 2019-04-13 22:55:55 INFO SparkEnv:54 - Registering MapOutput-
    Tracker
12. 2019-04-13 22:55:55 INFO SparkEnv:54 - Registering BlockMan-
    agerMaster
13. 2019-04-13 22:55:55 INFO BlockManagerMasterEndpoint:54 - U-
    sing org.apache.spark.storage.DefaultTopologyMapper for get-
    ting topology information
14. 2019-04-13 22:55:55 INFO BlockManagerMasterEndpoint:54 -
    BlockManagerMasterEndpoint up
15. 2019-04-13 22:55:55 INFO DiskBlockManager:54 - Created local
    directory at /tmp/blockmgr-6ad821a0-6411-49a5-a048-
    b625e8f33309
16. 2019-04-13 22:55:55 INFO MemoryStore:54 - MemoryStore started
    with capacity 366.3 MB
17. 2019-04-13 22:55:55 INFO SparkEnv:54 - Registering OutputCom-
    mitCoordinator
18. 2019-04-13 22:55:55 INFO log:192 - Logging initialized @1712ms
19. 2019-04-13 22:55:55 INFO Server:346 - jetty-9.3.z-SNAPSHOT
20. 2019-04-13 22:55:55 INFO Server:414 - Started @1803ms
21. 2019-04-13 22:55:55 INFO AbstractConnector:278 - Started
    ServerConnector@285f09de{HTTP/1.1,[http/1.1]}{0.0.0.0:4040}
22. 2019-04-13 22:55:55 INFO Utils:54 - Successfully started serv-
    ice 'SparkUI' on port 4040.
23. 2019-04-13 22:55:55 INFO ContextHandler:781 - Started o.s.j.s.
    ServletContextHandler@6e33c391{/jobs,null,AVAILABLE,@Spark}
24. 2019-04-13 22:55:55 INFO ContextHandler:781 - Started o.s.j.s.
    ServletContextHandler@32f61a31{/jobs/json,null,AVAILABLE,
    @Spark}
25. 2019-04-13 22:55:55 INFO ContextHandler:781 - Started o.s.j.s.
    ServletContextHandler@f5c79a6{/jobs/job,null,AVAILABLE,
    @Spark}
26. 2019-04-13 22:55:55 INFO ContextHandler:781 - Started o.s.j.s.
    ServletContextHandler@5305c37d{/jobs/job/json,null,AVAILA-
    BLE,@Spark}
27. 2019-04-13 22:55:55 INFO ContextHandler:781 - Started o.s.j.s.
    ServletContextHandler@51a06cbe{/stages,null,AVAILABLE,
    @Spark}
```

```
28. 2019-04-13 22:55:55 INFO ContextHandler:781 - Started o.s.j.s.
    ServletContextHandler@3dddbe65{/stages/json,null,AVAILABLE,
    @Spark}
29. 2019-04-13 22:55:55 INFO ContextHandler:781 - Started o.s.j.s.
    ServletContextHandler@49a64d82{/stages/stage,null,AVAILA-
    BLE,@Spark}
30. 2019-04-13 22:55:55 INFO ContextHandler:781 - Started o.s.j.s.
    ServletContextHandler@36ac8a63{/stages/stage/json,null,
    AVAILABLE,@Spark}
31. 2019-04-13 22:55:55 INFO ContextHandler:781 - Started o.s.j.s.
    ServletContextHandler@4d9d1b69{/stages/pool,null,AVAILA-
    BLE,@Spark}
32. 2019-04-13 22:55:55 INFO ContextHandler:781 - Started o.s.j.s.
    ServletContextHandler@52c8295b{/stages/pool/json,null,
    AVAILABLE,@Spark}
33. 2019-04-13 22:55:55 INFO ContextHandler:781 - Started o.s.j.s.
    ServletContextHandler@251f7d26{/storage,null,AVAILABLE,@
    Spark}
34. 2019-04-13 22:55:55 INFO ContextHandler:781 - Started o.s.j.s.
    ServletContextHandler@77b21474{/storage/json,null,AVAILA-
    BLE,@Spark}
35. 2019-04-13 22:55:55 INFO ContextHandler:781 - Started o.s.j.s.
    ServletContextHandler@52d10fb8{/storage/rdd,null,AVAILABLE,
    @Spark}
36. 2019-04-13 22:55:55 INFO ContextHandler:781 - Started o.s.j.s.
    ServletContextHandler@41c07648{/storage/rdd/json,null,
    AVAILABLE,@Spark}
37. 2019-04-13 22:55:55 INFO ContextHandler:781 - Started o.s.j.s.
    ServletContextHandler@1fe8d51b{/environment,null,AVAILABLE,
    @Spark}
38. 2019-04-13 22:55:55 INFO ContextHandler:781 - Started o.s.j.s.
    ServletContextHandler@781e7326{/environment/json,null,
    AVAILABLE,@Spark}
39. 2019-04-13 22:55:55 INFO ContextHandler:781 - Started o.s.j.s.
    ServletContextHandler@22680f52{/executors,null,AVAILABLE,
    @Spark}
40. 2019-04-13 22:55:55 INFO ContextHandler:781 - Started o.s.j.s.
```

```
    ServletContextHandler@60d84f61{/executors/json,null,AVAILA-
    BLE,@Spark}
41. 2019-04-13 22:55:55 INFO ContextHandler:781 - Started o.s.j.s.
    ServletContextHandler@39c11e6c{/executors/threadDump,null,
    AVAILABLE,@Spark}
42. 2019-04-13 22:55:55 INFO ContextHandler:781 - Started o.s.j.s.
    ServletContextHandler@324dcd31{/executors/threadDump/json,
    null,AVAILABLE,@Spark}
43. 2019-04-13 22:55:55 INFO ContextHandler:781 - Started o.s.j.s.
    ServletContextHandler@503d56b5{/static,null,AVAILABLE,
    @Spark}
44. 2019-04-13 22:55:55 INFO ContextHandler:781 - Started o.s.j.s.
    ServletContextHandler@7cd1ac19{/,null,AVAILABLE,@Spark}
45. 2019-04-13 22:55:55 INFO ContextHandler:781 - Started o.s.j.s.
    ServletContextHandler@2f40a43{/api,null,AVAILABLE,@Spark}
46. 2019-04-13 22:55:55 INFO ContextHandler:781 - Started o.s.j.s.
    ServletContextHandler@3a80515c{/jobs/job/kill,null,AVAILA-
    BLE,@Spark}
47. 2019-04-13 22:55:55 INFO ContextHandler:781 - Started o.s.j.s.
    ServletContextHandler@547e29a4{/stages/stage/kill,null,
    AVAILABLE,@Spark}
48. 2019-04-13 22:55:55 INFO SparkUI:54 - Bound SparkUI to 0.0.0.0,
    and started at http://mynode4:4040
49. 2019-04-13 22:55:55 INFO SparkContext:54 - Added JAR file:/
    software/spark-2.3.1/bin/../examples/jars/spark-examples_
    2.11-2.3.1.jar at spark://mynode4:50000/jars/spark-examples_
    2.11-2.3.1.jar with timestamp 1555167355771
50. 2019-04-13 22:55:55 INFO StandaloneAppClient$ClientEndpoint:
    54 - Connecting to master spark://mynode1:7077...
51. 2019-04-13 22:55:55 INFO TransportClientFactory:267 - Success-
    fully created connection to mynode1/192.168.179.13:7077 after
    67 ms (0 ms spent in bootstraps)
52. 2019-04-13 22:55:56 INFO StandaloneSchedulerBackend:54 - Con-
    nected to Spark cluster with app ID app-20190413225556-0001
53. 2019-04-13 22:55:56 INFO StandaloneAppClient$ClientEndpoint:
    54 - Executor added: app-20190413225556-0001/0 on worker-
```

```
    20190413212712 -192.168.179.15 -43616 (192.168.179.15:43616)
    with 8 core(s)
54.2019 -04 -13 22:55:56 INFO StandaloneSchedulerBackend:54 -Gran-
    ted executor ID app -20190413225556 -0001/0 on hostPort 192.
    168.179.15:43616 with 8 core(s), 1024.0 MB RAM
55.2019 -04 -13 22:55:56 INFO StandaloneAppClient$ClientEndpoint:
    54 -Executor added: app -20190413225556 -0001/1 on worker -
    20190413212712 -192.168.179.14 -43134 (192.168.179.14:43134)
    with 8 core(s)
56.2019 -04 -13 22:55:56 INFO StandaloneSchedulerBackend:54 -Gran-
    ted executor ID app -20190413225556 -0001/1 on hostPort 192.
    168.179.14:43134 with 8 core(s), 1024.0 MB RAM
57.2019 -04 -13 22:55:56 INFO Utils:54 -Successfully started serv-
    ice 'org.apache.spark.network.netty.NettyBlockTransferService'
    on port 53391.
58.2019 -04 -13 22:55:56 INFO NettyBlockTransferService:54 -Serv-
    er created on mynode4:53391
59.2019 -04 -13 22:55:56 INFO BlockManager:54 -Using org.apache.
    spark.storage.RandomBlockReplicationPolicy for block replica-
    tion policy
60.2019 -04 -13 22:55:56 INFO StandaloneAppClient$ClientEndpoint:
    54 -Executor updated: app -20190413225556 -0001/1 is now RUN-
    NING
61.2019 -04 -13 22:55:56 INFO StandaloneAppClient$ClientEndpoint:
    54 -Executor updated: app -20190413225556 -0001/0 is now RUN-
    NING
62.2019 -04 -13 22:55:56 INFO BlockManagerMaster:54 -Registering
    BlockManagerBlockManagerId(driver, mynode4, 53391, None)
63.2019 -04 -13 22:55:56 INFO BlockManagerMasterEndpoint:54 -Reg-
    istering block manager mynode4:53391 with 366.3 MB RAM, Block-
    ManagerId(driver, mynode4, 53391, None)
64.2019 -04 -13 22:55:56 INFO BlockManagerMaster:54 -Registered
    BlockManagerBlockManagerId(driver, mynode4, 53391, None)
65.2019 -04 -13 22:55:56 INFO BlockManager:54 -Initialized Block-
    Manager: BlockManagerId(driver, mynode4, 53391, None)
66.2019 -04 -13 22:55:56 INFO ContextHandler:781 -Started o.s.j.s.
    ServletContextHandler@6f099cef{/metrics/json,null,AVAILA-
    BLE,@Spark}
```

```
67. 2019 - 04 - 13 22:55:56 INFO StandaloneSchedulerBackend:54 - SchedulerBackend is ready for scheduling beginning after reached minRegisteredResourcesRatio: 0.0
68. 2019 - 04 - 13 22:55:57 INFO SparkContext:54 - Starting job: reduce at SparkPi.scala:38
69. 2019 - 04 - 13 22:55:57 INFO DAGScheduler:54 - Got job 0 (reduce at SparkPi.scala:38) with 2 output partitions
70. 2019 - 04 - 13 22:55:57 INFO DAGScheduler:54 - Final stage: ResultStage 0 (reduce at SparkPi.scala:38)
71. 2019 - 04 - 13 22:55:57 INFO DAGScheduler:54 - Parents of final stage: List()
72. 2019 - 04 - 13 22:55:57 INFO DAGScheduler:54 - Missing parents: List()
73. 2019 - 04 - 13 22:55:57 INFO DAGScheduler:54 - Submitting ResultStage 0 (MapPartitionsRDD[1] at map at SparkPi.scala:34), which has no missing parents
74. 2019 - 04 - 13 22:55:57 INFO MemoryStore:54 - Block broadcast_0 stored as values in memory (estimated size 1832.0 B, free 366.3 MB)
75. 2019 - 04 - 13 22:55:57 INFO MemoryStore:54 - Block broadcast_0_piece0 stored as bytes in memory (estimated size 1181.0 B, free 366.3 MB)
76. 2019 - 04 - 13 22:55:57 INFO BlockManagerInfo:54 - Added broadcast_0_piece0 in memory on mynode4:53391 (size: 1181.0 B, free: 366.3 MB)
77. 2019 - 04 - 13 22:55:57 INFO SparkContext:54 - Created broadcast 0 from broadcast at DAGScheduler.scala:1039
78. 2019 - 04 - 13 22:55:57 INFO DAGScheduler:54 - Submitting 2 missing tasks from ResultStage 0 (MapPartitionsRDD[1] at map at SparkPi.scala:34) (first 15 tasks are for partitions Vector(0, 1))
79. 2019 - 04 - 13 22:55:57 INFO TaskSchedulerImpl:54 - Adding task set 0.0 with 2 tasks
80. 2019 - 04 - 13 22:55:58 INFO CoarseGrainedSchedulerBackend$DriverEndpoint:54 - Registered executor NettyRpcEndpointRef(spark - client://Executor) (192.168.179.15:35726) with ID 0
81. 2019 - 04 - 13 22:55:58 INFO TaskSetManager:54 - Starting task 0.0 in stage 0.0 (TID 0, 192.168.179.15, executor 0, partition 0, PROCESS_LOCAL, 7857 bytes)
```

```
2019-04-13 22:55:58 INFO TaskSetManager:54 - Starting task 1.0 in stage 0.0 (TID 1, 192.168.179.15, executor 0, partition 1, PROCESS_LOCAL, 7857 bytes)
2019-04-13 22:55:58 INFO CoarseGrainedSchedulerBackend$DriverEndpoint:54 - Registered executor NettyRpcEndpointRef(spark-client://Executor) (192.168.179.14:46612) with ID 1
2019-04-13 22:55:58 INFO BlockManagerMasterEndpoint:54 - Registering block manager 192.168.179.14:36939 with 366.3 MB RAM, BlockManagerId(1, 192.168.179.14, 36939, None)
2019-04-13 22:55:58 INFO BlockManagerMasterEndpoint:54 - Registering block manager 192.168.179.15:52570 with 366.3 MB RAM, BlockManagerId(0, 192.168.179.15, 52570, None)
2019-04-13 22:55:59 INFO BlockManagerInfo:54 - Added broadcast_0_piece0 in memory on 192.168.179.15:52570 (size: 1181.0 B, free: 366.3 MB)
2019-04-13 22:55:59 INFO TaskSetManager:54 - Finished task 1.0 in stage 0.0 (TID 1) in 714 ms on 192.168.179.15 (executor 0) (1/2)
2019-04-13 22:55:59 INFO TaskSetManager:54 - Finished task 0.0 in stage 0.0 (TID 0) in 749 ms on 192.168.179.15 (executor 0) (2/2)
2019-04-13 22:55:59 INFO TaskSchedulerImpl:54 - Removed TaskSet 0.0, whose tasks have all completed, from pool
2019-04-13 22:55:59 INFO DAGScheduler:54 - ResultStage 0 (reduce at SparkPi.scala:38) finished in 1.928 s
2019-04-13 22:55:59 INFO DAGScheduler:54 - Job 0 finished: reduce at SparkPi.scala:38, took 2.045323 s
Pi is roughly 3.1427157135785677 #执行结果
2019-04-13 22:55:59 INFO AbstractConnector:318 - Stopped Spark@285f09de{HTTP/1.1,[http/1.1]}{0.0.0.0:4040}
2019-04-13 22:55:59 INFO SparkUI:54 - Stopped Spark web UI at http://mynode4:4040
2019-04-13 22:55:59 INFO StandaloneSchedulerBackend:54 - Shutting down all executors
2019-04-13 22:55:59 INFO CoarseGrainedSchedulerBackend$DriverEndpoint:54 - Asking each executor to shut down
2019-04-13 22:55:59 INFO MapOutputTrackerMasterEndpoint:54 - MapOutputTrackerMasterEndpoint stopped!
2019-04-13 22:55:59 INFO MemoryStore:54 - MemoryStore cleared
```

```
2019-04-13 22:55:59 INFO BlockManager:54 - BlockManager stopped
2019-04-13 22:55:59 INFO BlockManagerMaster:54 - BlockManagerMaster stopped
2019-04-13 22:55:59 INFO OutputCommitCoordinator$OutputCommitCoordinatorEndpoint:54 - OutputCommitCoordinator stopped!
2019-04-13 22:55:59 INFO SparkContext:54 - Successfully stopped SparkContext
2019-04-13 22:55:59 INFO ShutdownHookManager:54 - Shutdown hook called
2019-04-13 22:55:59 INFO ShutdownHookManager:54 - Deleting directory /tmp/spark-af8ee0c5-8834-4731-acde-fc9dd44eea8c
2019-04-13 22:55:59 INFO ShutdownHookManager:54 - Deleting directory /tmp/spark-d1d1f274-1214-49fd-bb6a-713ff4bca504
```

经过以上的运行，发现在 Spark 客户端上提交任务也是没有问题的。此时，Spark 集群系统搭建完成，可以将 ×× 系统运行在 Spark 集群上进行工作。

# 项目三
# Spark 基于 Yarn 搭建

【项目描述】

上一项目介绍了 Spark 集群的搭建，实际上搭建的就是 Spark 的 Standalone 集群，也就是说，提交 Spark 任务运行所依赖的资源调度是 Spark 集群，任务就是运行在 Spark 集群中的 Worker 节点上。

【项目分析】

在 Hadoop 生态圈中 Hadoop2. x 版本之后引入了 Yarn，引入 Yarn 的目的是将 MapReduce 任务的资源调度和任务调度分离，Yarn 只管理资源调度，任务调度由 ApplicationMaster 来管理。MapReduce 任务可以基于 Yarn 资源调度框架进行调度，Spark 任务同样也可以基于 Yarn 资源调度框架进行调度。

Spark 基于 Yarn 进行资源调度，还需要做一些简单的配置。下面详细介绍 Spark 任务基于 Yarn 调度的基础条件、搭建过程及任务提交。

## 任务 3.1　基于 Yarn 构建介绍

### 3.1.1　基于 Yarn 搭建环境配置

Spark 任务基于 Yarn 资源调度框架进行调度时，首要条件是搭建好Hadoop 集群环境。Hadoop 集群环境中重要的角色如下。

**1. NameNode**

NameNode 管理 HDFS 文件系统的命名空间，它维护着文件系统及文件系统内所有的文件和目录，这些信息以两个文件形式永久保存在本地磁盘上：命名空间镜像文件和编辑日志文件。NameNode 也记录着每个文件中各个数据块所在的数据节点信息，但它并不永久保存数据块的位置信息，因为这些信息会在系统启动时根据数据节点信息重建。

**2. DataNode**

DataNode 是 HDFS 文件系统的工作节点，它们根据需要存储并检索数据块（受 NameNode

调度)，并且定时向 NameNode 发送它们存储的数据块的列表。

**3. ResourceManager**

ResourceManager 是管理集群所有可用资源的中心节点，并能够帮助管理 Yarn 上的分部署 applications。它与每个节点上的 NodeManagers（NMs）和 ApplicationMasters（AMs）一起工作。

**4. NodeManager**

NodeManager 是 Yarn 中每个节点上的代理，它管理 Hadoop 集群中单个计算节点，包括与 ResourceManger 保持通信、监督 Container 的生命周期管理、监控每个 Container 的资源使用（内存、CPU 等）情况、追踪节点健康状况、管理日志和不同应用程序用到的附属服务（auxiliary service）等。

表 3－1 列举了 Hadoop 集群节点以上角色的详细划分。

表 3－1 Hadoop 集群节点角色的详细划分

| 角色 | mynode1 | mynode2 | mynode3 | mynode4 | mynode5 |
|---|---|---|---|---|---|
| NameNode | ● | ● |  | × | × |
| DataNode | × | × | ● | ● | ● |
| ResourceManager | ● | ● | × | × | × |
| NodeManager | × | × | ● | ● | ● |

### 3.1.2 Hadoop 集群

选择完全分布式搭建 Hadoop 集群，需要依赖 ZooKeeper 分布式协调系统进行协调管理，因此，需要安装 ZooKeeper 集群。ZooKeeper 集群的角色节点划分见表 3－2。

表 3－2 角色节点划分

| 角色 | mynode1 | mynode2 | mynode3 | mynode4 | mynode5 |
|---|---|---|---|---|---|
| ZooKeeper | × | × | ● | ● | ● |

ZooKeeper 搭建过程如下。

登录 ZooKeeper 官网，下载 ZooKeeper，这里选择的版本是 zookeeper－3.4.5，下载地址如下：

http://archive.apache.org/dist/zookeeper/zookeeper-3.4.5/

将下载好的 zookeeper－3.4.5 上传至 mynode3 节点/software 路径下，进行解压：

```
5.[root@mynode3 software]# tar -zxvf ./zookeeper-3.4.5.tar.gz
```

在 mynode3 节点上配置 ZooKeeper 的环境变量，操作如下：

```
#编辑 /etc/profile 文件,在文件最后追加如下内容,并保存
[root@mynode3 software]# vim /etc/profile
export ZOOKEEPER_HOME = /software/zookeeper -3.4.5
export PATH =$PATH: $ZOOKEEPER_HOME/bin

#使 profile 文件配置生效
[root@mynode3 software]# source /etc/profile
```

在 ZooKeeper 路径/software/zookeeper-3.4.5/conf 下配置 zoo.cfg 文件，操作如下：

```
#进入对应路径,修改 zoo.cfg 名称
[root@mynode3 conf]# cd /software/zookeeper -3.4.5/conf/
[root@mynode3 conf]# mv zoo_sample.cfgzoo.cfg

#编辑 zoo.cfg 文件
[root@mynode3 conf]# vim zoo.cfg

#编辑内容如下:
tickTime =2000
dataDir = /opt/data/zookeeper
clientPort =2181
initLimit =5
syncLimit =2
server.1 =mynode3:2888:3888
server.2 =mynode4:2888:3888
server.3 =mynode5:2888:3888
```

将以上配置好的信息发送到 mynode4、mynode5 节点，并按照上面的方式配置 ZooKeeper 的环境变量：

```
[root@mynode3 zookeeper -3.4.5]# cd /software/
[root@mynode3 software]#scp -r ./zookeeper -3.4.5 root@mynode4:'pwd'
[root@mynode3 software]#scp -r ./zookeeper -3.4.5 root@mynode5:'pwd'
```

在 mynode3、mynode4、mynode5 节点上分别创建路径/opt/data/zookeeper 存放 ZooKeeper 数据：

```
#在 mynode3 上创建路径
[root@mynode3 software]#mkdir -p /opt/data/zookeeper

```

```
#在 mynode4 上创建路径
[root@mynode4 software]#mkdir -p /opt/data/zookeeper

#在 mynode5 上创建路径
[root@mynode5 software]#mkdir -p /opt/data/zookeeper
```

在 mynode3、mynode4、mynode5 三个节点创建节点 ID，在每台节点配置的 dataDir 路径/opt/data/zookeeper 中添加 myid 文件，操作如下：

```
#在 mynode3 上路径/opt/data/zookeeper 中创建 myid 文件,并写入1
[root@mynode3 software]# cd /opt/data/zookeeper/
[root@mynode3 zookeeper]# vim myid #写入1 并保存即可

#在 mynode4 上路径/opt/data/zookeeper 中创建 myid 文件,并写入2
[root@mynode4 software]# cd /opt/data/zookeeper/
[root@mynode4 zookeeper]# vim myid #写入2 并保存即可

#在 mynode5 上路径/opt/data/zookeeper 中创建 myid 文件,并写入3
[root@mynode5 software]# cd /opt/data/zookeeper/
[root@mynode5 zookeeper]# vim myid #写入3 并保存即可
```

在每台节点中启动 ZooKeeper 集群并验证每台节点 ZooKeeper 的角色，验证 ZooKeeper 是否安装成功：

```
#在 mynode3 启动 ZooKeeper
[root@mynode3 zookeeper]# zkServer.sh start
JMX enabled by default
Using config: /software/zookeeper-3.4.5/bin/../conf/zoo.cfg
Starting zookeeper ... STARTED

#在 mynode4 启动 ZooKeeper
[root@mynode4 zookeeper]# zkServer.sh start
JMX enabled by default
Using config: /software/zookeeper-3.4.5/bin/../conf/zoo.cfg
Starting zookeeper ... STARTED

#在 mynode5 启动 ZooKeeper
[root@mynode5 zookeeper]# zkServer.sh start
JMX enabled by default
Using config: /software/zookeeper-3.4.5/bin/../conf/zoo.cfg
```

```
Starting zookeeper ... STARTED

#在 mynode3 验证 ZooKeeper 的状态
[root@mynode3 zookeeper]# zkServer.sh status
JMX enabled by default
Using config: /software/zookeeper-3.4.5/bin/../conf/zoo.cfg
Mode: follower

#在 mynode4 验证 ZooKeeper 的状态
[root@mynode4 zookeeper]# zkServer.sh status
JMX enabled by default
Using config: /software/zookeeper-3.4.5/bin/../conf/zoo.cfg
Mode: leader

#在 mynode5 验证 ZooKeeper 的状态
[root@mynode5 zookeeper]# zkServer.sh status
JMX enabled by default
Using config: /software/zookeeper-3.4.5/bin/../conf/zoo.cfg
Mode: follower
```

完成上述步骤后，ZooKeeper 安装成功，下面安装 Hadoop 完全分布式集群。

在安装好 ZooKeeper 系统的集群上搭建 Hadoop。首先去 Hadoop 的官网下载 hadoop-2.6.5，下载地址如下：

https://archive.apache.org/dist/hadoop/common/hadoop-2.6.5/

将下载好的 hadoop-2.6.5.tar.gz 压缩包上传至 mynod1 节点/software 路径下，解压：

```
[root@mynode1 software]# tar -zxvf ./hadoop-2.6.5.tar.gz
```

解压完成之后，可以配置 Hadoop 的环境变量，配置如下。

打开/etc/profile 配置 Hadoop 的环境变量：

```
[root@mynode1 software]# vim /etc/profile
#在文件最后追加如下内容:
export HADOOP_HOME=/software/hadoop-2.6.5
export PATH=$PATH:$HADOOP_HOME/bin:$HADOOP_HOME/sbin:

#最后保存文件,使文件生效
[root@mynode1 software]# source /etc/profile
```

在/software/hadoop-2.6.5/etc/hadoop 路径中的 hadoop-env.sh 中加入 JAVA_HOME：

```
[root@mynode1 hadoop]# vim /software/hadoop-2.6.5/etc/hadoop/hadoop-env.sh
#加入 JAVA_HOME,并保存
export JAVA_HOME=/software/jdk1.8.0_181
```

配置/software/hadoop-2.6.5/etc/hadoop/hdfs-site.xml 文件，配置内容如下：

```
<?xml version="1.0" encoding="UTF-8"?>
<?xml-stylesheet type="text/xsl" href="configuration.xsl"?>
<!--
Licensed under the Apache License, Version 2.0 (the "License");
you may not use this file except in compliance with the License.
You may obtain a copy of the License at

http://www.apache.org/licenses/LICENSE-2.0

Unless required by applicable law or agreed to in writing, software
distributed under the License is distributed on an "AS IS" BASIS,
WITHOUT WARRANTIES OR CONDITIONS OF ANY KIND, either express or implied.
See the License for the specific language governing permissions and
limitations under the License. See accompanying LICENSE file.
-->

<!-- Put site-specific property overrides in this file. -->

<configuration>
<property>
<!--这里配置集群逻辑名称,可以随意写 -->
<name>dfs.nameservices</name>
<value>mycluster</value>
</property>
<property>
<!--禁用权限 -->
<name>dfs.permissions.enabled</name>
<value>false</value>
```

```
</property>
<property>
<!--配置 namenode 的名称,多个用逗号分割 -->
<name>dfs.ha.namenodes.mycluster</name>
<value>nn1,nn2</value>
</property>
<property>
<!--dfs.namenode.rpc-address.[nameservice ID].[name node ID] namenode 所在服务器名称和 RPC 监听端口号 -->
<name>dfs.namenode.rpc-address.mycluster.nn1</name>
<value>mynode1:8020</value>
</property>
<property>
<!--dfs.namenode.rpc-address.[nameservice ID].[name node ID] namenode 所在服务器名称和 RPC 监听端口号 -->
<name>dfs.namenode.rpc-address.mycluster.nn2</name>
<value>mynode2:8020</value>
</property>
<property>
<!--dfs.namenode.http-address.[nameservice ID].[name node ID] namenode 监听的 HTTP 协议端口 -->
<name>dfs.namenode.http-address.mycluster.nn1</name>
<value>mynode1:50070</value>
</property>
<property>
<!--dfs.namenode.http-address.[nameservice ID].[name node ID] namenode 监听的 HTTP 协议端口 -->
<name>dfs.namenode.http-address.mycluster.nn2</name>
<value>mynode2:50070</value>
</property>

<property>
<!--namenode 共享的编辑目录,journalnode 所在服务器名称和监听的端口 -->
<name>dfs.namenode.shared.edits.dir</name>
<value>qjournal://mynode3:8485;mynode4:8485;mynode5:8485/mycluster</value>
```

```
</property>

<property>
<!--namenode 高可用代理类 -->
<name>dfs.client.failover.proxy.provider.mycluster</name>
<value>org.apache.hadoop.hdfs.server.namenode.ha.ConfiguredFailoverProxyProvider</value>
</property>

<property>
<!--使用 ssh 免密码自动登录 -->
<name>dfs.ha.fencing.methods</name>
<value>sshfence</value>
</property>

<property>
<name>dfs.ha.fencing.ssh.private-key-files</name>
<value>/root/.ssh/id_rsa</value>
</property>
<property>
<!--journalnode 存储数据的地方 -->
<name>dfs.journalnode.edits.dir</name>
<value>/opt/data/journal/node/local/data</value>
</property>

<property>
<!--配置 namenode 自动切换 -->
<name>dfs.ha.automatic-failover.enabled</name>
<value>true</value>
</property>
</configuration>
```

配置/software/hadoop-2.6.5/etc/hadoop/core-site.xml 文件，配置内容如下：

```
<?xml version="1.0" encoding="UTF-8"?>
<?xml-stylesheet type="text/xsl" href="configuration.xsl"?>
<!--
Licensed under the Apache License, Version 2.0 (the "License");
```

```
you may not use this file except in compliance with the License.
You may obtain a copy of the License at

http://www.apache.org/licenses/LICENSE-2.0

Unless required by applicable law or agreed to in writing, software
distributed under the License is distributed on an "AS IS" BASIS,
WITHOUT WARRANTIES OR CONDITIONS OF ANY KIND, either express or implied.
See the License for the specific language governing permissions and
limitations under the License. See accompanying LICENSE file.
-->

<!-- Put site-specific property overrides in this file. -->

<configuration>
<property>
<!--为 Hadoop 客户端配置默认的高可用路径 -->
<name>fs.defaultFS</name>
<value>hdfs://mycluster</value>
</property>
<property>
<!-- Hadoop 数据存放的路径,namenode,datanode 数据存放路径都依赖本路径,不要使用 file:/开头,使用绝对路径即可
namenode 默认存放路径:file://${hadoop.tmp.dir}/dfs/name
datanode 默认存放路径:file://${hadoop.tmp.dir}/dfs/data
-->
<name>hadoop.tmp.dir</name>
<value>/opt/data/hadoop/</value>
</property>

<property>
<!--指定 zookeeper 所在的节点 -->
<name>ha.zookeeper.quorum</name>
```

```
<value>mynode3:2181,mynode4:2181,mynode5:2181</value>
</property>
</configuration>
```

配置/software/hadoop-2.6.5/etc/hadoop/yarn-site.xml 文件，配置内容如下：

```
<?xml version="1.0"?>
<!--
Licensed under the Apache License, Version 2.0 (the "License");
you may not use this file except in compliance with the License.
You may obtain a copy of the License at

http://www.apache.org/licenses/LICENSE-2.0

Unless required by applicable law or agreed to in writing, software
distributed under the License is distributed on an "AS IS" BASIS,
WITHOUT WARRANTIES OR CONDITIONS OF ANY KIND, either express or implied.
See the License for the specific language governing permissions and
limitations under the License. See accompanying LICENSE file.
-->
<configuration>

<!-- Site specific YARN configuration properties -->
<property>
<name>yarn.nodemanager.aux-services</name>
<value>mapreduce_shuffle</value>
</property>
<property>
<name>yarn.nodemanager.env-whitelist</name>
<value>JAVA_HOME,HADOOP_COMMON_HOME,HADOOP_HDFS_HOME,HADOOP_CONF_DIR,CLASSPATH_PREPEND_DISTCACHE,HADOOP_YARN_HOME,HADOOP_MAPRED_HOME</value>
</property>

<property>
<!--配置 Yarn 为高可用 -->
```

```
<name>yarn.resourcemanager.ha.enabled</name>
<value>true</value>
</property>
<property>
<!--集群的唯一标识 -->
<name>yarn.resourcemanager.cluster-id</name>
<value>mycluster</value>
</property>
<property>
<!--ResourceManager ID -->
<name>yarn.resourcemanager.ha.rm-ids</name>
<value>rm1,rm2</value>
</property>
<property>
<!--指定 ResourceManager 所在的节点 -->
<name>yarn.resourcemanager.hostname.rm1</name>
<value>mynode1</value>
</property>
<property>
<!--指定 ResourceManager 所在的节点 -->
<name>yarn.resourcemanager.hostname.rm2</name>
<value>mynode2</value>
</property>
<property>
<!--指定 ResourceManager Http 监听的节点 -->
<name>yarn.resourcemanager.webapp.address.rm1</name>
<value>mynode1:8088</value>
</property>
<property>
<!--指定 ResourceManager Http 监听的节点 -->
<name>yarn.resourcemanager.webapp.address.rm2</name>
<value>mynode2:8088</value>
</property>
<property>
<!--指定 ZooKeeper 所在的节点 -->
<name>yarn.resourcemanager.zk-address</name>
<value>mynode3:2181,mynode4:2181,mynode5:2181</value>
```

```
</property>
</configuration>
```

配置/software/hadoop-2.6.5/etc/hadoop/map-site.xml 文件，配置内容如下：

```
[root@mynode1 hadoop]# cp /software/hadoop-2.6.5/etc/hadoop/mapred-site.xml.template /software/hadoop-2.6.5/etc/hadoop/mapred-site.xml
[root@mynode1 hadoop]# vim /software/hadoop-2.6.5/etc/hadoop/mapred-site.xml

#mapred-site.xml 配置信息如下:
<?xml version="1.0"?>
<?xml-stylesheet type="text/xsl" href="configuration.xsl"?>
<!--
Licensed under the Apache License, Version 2.0 (the "License");
you may not use this file except in compliance with the License.
You may obtain a copy of the License at

http://www.apache.org/licenses/LICENSE-2.0

Unless required by applicable law or agreed to in writing, software
distributed under the License is distributed on an "AS IS" BASIS,
WITHOUT WARRANTIES OR CONDITIONS OF ANY KIND, either express or implied.
See the License for the specific language governing permissions and
limitations under the License. See accompanying LICENSE file.
-->

<!-- Put site-specific property overrides in this file. -->

<configuration>
<property>
<name>mapreduce.framework.name</name>
<value>yarn</value>
```

```
</property>
</configuration>
```

配置/software/hadoop-2.6.5/etc/hadoop/slaves 文件，配置 DataNode 节点信息，配置信息如下：

```
mynode3
mynode4
mynode5
```

至此，Hadoop 的配置已经完成，现在将/software/hadoop-2.6.5 目录发送到 mynode2、mynode3、mynode4、mynode5 节点上，同时，在这些节点上配置 Hadoop 的环境变量，并使环境变量生效。

```
[root@mynode1 software]# cd /software/
[root@mynode1 software]# scp -r ./hadoop-2.6.5 mynode2:'pwd'
[root@mynode1 software]# scp -r ./hadoop-2.6.5 mynode3:'pwd'
[root@mynode1 software]# scp -r ./hadoop-2.6.5 mynode4:'pwd'
[root@mynode1 software]# scp -r ./hadoop-2.6.5 mynode5:'pwd'
```

此时，Hadoop 的配置在各个节点上已经全部完成，下面进行 Hadoop 的格式化及启动。

- 启动 ZooKeeper：

```
#在 mynode3 上启动 ZooKeeper
[root@mynode3 ~]# zkServer.sh start
JMX enabled by default
Using config: /software/zookeeper-3.4.5/bin/../conf/zoo.cfg
Starting zookeeper ... STARTED

#在 mynode4 上启动 ZooKeeper
[root@mynode4 ~]# zkServer.sh start
JMX enabled by default
Using config: /software/zookeeper-3.4.5/bin/../conf/zoo.cfg
Starting zookeeper ... STARTED

#在 mynode5 上启动 ZooKeeper
[root@mynode5 ~]# zkServer.sh start
JMX enabled by default
Using config: /software/zookeeper-3.4.5/bin/../conf/zoo.cfg
Starting zookeeper ... STARTED
```

- 在 mynode1 的 NameNode 中格式化 ZooKeeper：

```
[root@mynode1 software]#hdfszkfc -formatZK
19/04/14 22:11:14 WARN util.NativeCodeLoader: Unable to load native-hadoop library for your platform... using builtin-java classes where applicable
19/04/14 22:11:14 INFO tools.DFSZKFailoverController: Failover controller configured for NameNodeNameNode at mynode1/192.168.179.13:8020
19/04/14 22:11:14 INFO zookeeper.ZooKeeper: Client environment:zookeeper.version=3.4.6-1569965, built on 02/20/2014 09:09 GMT
19/04/14 22:11:14 INFO zookeeper.ZooKeeper: Client environment:host.name=mynode1
19/04/14 22:11:14 INFO zookeeper.ZooKeeper: Client environment:java.version=1.8.0_181
19/04/14 22:11:14 INFO zookeeper.ZooKeeper: Client environment:java.vendor=Oracle Corporation
19/04/14 22:11:14 INFO zookeeper.ZooKeeper: Client environment:java.home=/software/jdk1.8.0_181/jre
19/04/14 22:11:14 INFO zookeeper.ZooKeeper: Client environment:java.class.path=/software/hadoop-2.6.5/etc/hadoop:/software/hadoop-2.6.5/share/hadoop/common/lib/avro-1.7.4.jar:/software/hadoop-
2.6.5/share/hadoop/common/lib/commons-lang-2.6.jar:/software/hadoop-
   2.6.5/share/hadoop/common/lib/jackson-core-asl-1.9.13.jar:/software/hadoop-
   2.6.5/share/hadoop/common/lib/commons-codec-1.4.jar:/software/hadoop-
   2.6.5/share/hadoop/common/lib/jaxb-api-2.2.2.jar:/software/hadoop-
   2.6.5/share/hadoop/common/lib/commons-cli-1.2.jar:/software/hadoop-
   2.6.5/share/hadoop/common/lib/commons-configuration-1.6.jar:/software/hadoop-
   2.6.5/share/hadoop/common/lib/jersey-json-1.9.jar:/software/hadoop-
   2.6.5/share/hadoop/common/lib/commons-el-1.0.jar:/software/hadoop-
```

```
2.6.5/share/hadoop/common/lib/mockito-all-1.8.5.jar:/software/hadoop-
2.6.5/share/hadoop/common/lib/htrace-core-3.0.4.jar:/software/hadoop-
2.6.5/share/hadoop/common/lib/xz-1.0.jar:/software/hadoop-
2.6.5/share/hadoop/common/lib/curator-framework-2.6.0.jar:/software/hadoop-
2.6.5/share/hadoop/common/lib/zookeeper-3.4.6.jar:/software/hadoop-
2.6.5/share/hadoop/common/lib/paranamer-2.3.jar:/software/hadoop-
2.6.5/share/hadoop/common/lib/jersey-server-1.9.jar:/software/hadoop-
2.6.5/share/hadoop/common/lib/commons-logging-1.1.3.jar:/software/hadoop-
2.6.5/share/hadoop/common/lib/servlet-api-2.5.jar:/software/hadoop-
2.6.5/share/hadoop/common/lib/apacheds-kerberos-codec-2.0.0-M15.jar:/software/hadoop-
2.6.5/share/hadoop/common/lib/commons-net-3.1.jar:/software/hadoop-
2.6.5/share/hadoop/common/lib/netty-3.6.2.Final.jar:/software/hadoop-
2.6.5/share/hadoop/common/lib/commons-compress-1.4.1.jar:/software/hadoop-
2.6.5/share/hadoop/common/lib/protobuf-java-2.5.0.jar:/software/hadoop-
2.6.5/share/hadoop/common/lib/snappy-java-1.0.4.1.jar:/software/hadoop-
2.6.5/share/hadoop/common/lib/api-util-1.0.0-M20.jar:/software/hadoop-
2.6.5/share/hadoop/common/lib/commons-math3-3.1.1.jar:/software/hadoop-
2.6.5/share/hadoop/common/lib/hamcrest-core-1.3.jar:/software/hadoop-
2.6.5/share/hadoop/common/lib/commons-digester-1.8.jar:/software/hadoop-
```

```
2.6.5/share/hadoop/common/lib/gson-2.2.4.jar:/software/ha-
doop-
2.6.5/share/hadoop/common/lib/activation-1.1.jar:/soft-
ware/hadoop-
2.6.5/share/hadoop/common/lib/xmlenc-0.52.jar:/software/
hadoop-
2.6.5/share/hadoop/common/lib/jetty-util-6.1.26.jar:/soft-
ware/hadoop-
2.6.5/share/hadoop/common/lib/java-xmlbuilder-0.4.jar:/
software/hadoop-
2.6.5/share/hadoop/common/lib/hadoop-auth-2.6.5.jar:/soft-
ware/hadoop-
2.6.5/share/hadoop/common/lib/jackson-xc-1.9.13.jar:/soft-
ware/hadoop-
2.6.5/share/hadoop/common/lib/httpclient-4.2.5.jar:/soft-
ware/hadoop-
2.6.5/share/hadoop/common/lib/jsr305-1.3.9.jar:/software/
hadoop-
2.6.5/share/hadoop/common/lib/jasper-runtime-5.5.23.jar:/
software/hadoop-
2.6.5/share/hadoop/common/lib/commons-beanutils-1.7.0.
jar:/software/hadoop-
2.6.5/share/hadoop/common/lib/apacheds-i18n-2.0.0-M15.
jar:/software/hadoop-
2.6.5/share/hadoop/common/lib/jaxb-impl-2.2.3-1.jar:/
software/hadoop-
2.6.5/share/hadoop/common/lib/jsch-0.1.42.jar:/software/
hadoop-
2.6.5/share/hadoop/common/lib/guava-11.0.2.jar:/software/
hadoop-
2.6.5/share/hadoop/common/lib/curator-client-2.6.0.jar:/
software/hadoop-
2.6.5/share/hadoop/common/lib/hadoop-annotations-2.6.5.
jar:/software/hadoop-
2.6.5/share/hadoop/common/lib/asm-3.2.jar:/software/hadoop-
2.6.5/share/hadoop/common/lib/commons-io-2.4.jar:/soft-
ware/hadoop-
```

```
2.6.5/share/hadoop/common/lib/junit-4.11.jar:/software/ha-
doop-
2.6.5/share/hadoop/common/lib/jackson-mapper-asl-1.9.13.
jar:/software/hadoop-
2.6.5/share/hadoop/common/lib/curator-recipes-2.6.0.jar:/
software/hadoop-
2.6.5/share/hadoop/common/lib/commons-httpclient-3.1.jar:/
software/hadoop-
2.6.5/share/hadoop/common/lib/httpcore-4.2.5.jar:/soft-
ware/hadoop-
2.6.5/share/hadoop/common/lib/api-asn1-api-1.0.0-M20.
jar:/software/hadoop-
2.6.5/share/hadoop/common/lib/slf4j-api-1.7.5.jar:/soft-
ware/hadoop-
2.6.5/share/hadoop/common/lib/slf4j-log4j12-1.7.5.jar:/
software/hadoop-
2.6.5/share/hadoop/common/lib/jettison-1.1.jar:/software/
hadoop-
2.6.5/share/hadoop/common/lib/jasper-compiler-5.5.23.jar:/
software/hadoop-
2.6.5/share/hadoop/common/lib/stax-api-1.0-2.jar:/soft-
ware/hadoop-
2.6.5/share/hadoop/common/lib/jets3t-0.9.0.jar:/software/
hadoop-
2.6.5/share/hadoop/common/lib/jersey-core-1.9.jar:/soft-
ware/hadoop-
2.6.5/share/hadoop/common/lib/jackson-jaxrs-1.9.13.jar:/
software/hadoop-
2.6.5/share/hadoop/common/lib/commons-beanutils-core-1.8.
0.jar:/software/hadoop-
2.6.5/share/hadoop/common/lib/jetty-6.1.26.jar:/software/
hadoop-
2.6.5/share/hadoop/common/lib/log4j-1.2.17.jar:/software/
hadoop-
2.6.5/share/hadoop/common/lib/jsp-api-2.1.jar:/software/
hadoop-
2.6.5/share/hadoop/common/lib/commons-collections-3.2.2.
jar:/software/hadoop-
```

```
2.6.5/share/hadoop/common/hadoop-common-2.6.5-tests.jar:/
software/hadoop-
2.6.5/share/hadoop/common/hadoop-common-2.6.5.jar:/soft-
ware/hadoop-
2.6.5/share/hadoop/common/hadoop-nfs-2.6.5.jar:/software/
hadoop-
2.6.5/share/hadoop/hdfs:/software/hadoop-2.6.5/share/ha-
doop/hdfs/lib/commons-lang-
2.6.jar:/software/hadoop-2.6.5/share/hadoop/hdfs/lib/jack-
son-core-asl-
1.9.13.jar:/software/hadoop-2.6.5/share/hadoop/hdfs/lib/
commons-codec-
1.4.jar:/software/hadoop-2.6.5/share/hadoop/hdfs/lib/com-
mons-cli-
1.2.jar:/software/hadoop-2.6.5/share/hadoop/hdfs/lib/xml-
apis-
1.3.04.jar:/software/hadoop-2.6.5/share/hadoop/hdfs/lib/
commons-el-
1.0.jar:/software/hadoop-2.6.5/share/hadoop/hdfs/lib/
htrace-core-
3.0.4.jar:/software/hadoop-2.6.5/share/hadoop/hdfs/lib/
jersey-server-
1.9.jar:/software/hadoop-2.6.5/share/hadoop/hdfs/lib/com-
mons-logging-
1.1.3.jar:/software/hadoop-2.6.5/share/hadoop/hdfs/lib/
servlet-api-
2.5.jar:/software/hadoop-2.6.5/share/hadoop/hdfs/lib/com-
mons-daemon-
1.0.13.jar:/software/hadoop-2.6.5/share/hadoop/hdfs/lib/
netty-
3.6.2.Final.jar:/software/hadoop-2.6.5/share/hadoop/hdfs/
lib/protobuf-java-
2.5.0.jar:/software/hadoop-2.6.5/share/hadoop/hdfs/lib/xm-
lenc-
0.52.jar:/software/hadoop-2.6.5/share/hadoop/hdfs/lib/jet-
ty-util-
6.1.26.jar:/software/hadoop-2.6.5/share/hadoop/hdfs/lib/
jsr305-
```

```
1.3.9.jar:/software/hadoop-2.6.5/share/hadoop/hdfs/lib/
jasper-runtime-
5.5.23.jar:/software/hadoop-2.6.5/share/hadoop/hdfs/lib/
guava-
11.0.2.jar:/software/hadoop-2.6.5/share/hadoop/hdfs/lib/
asm-3.2.jar:/software/hadoop-
2.6.5/share/hadoop/hdfs/lib/commons-io-2.4.jar:/software/
hadoop-
2.6.5/share/hadoop/hdfs/lib/jackson-mapper-asl-1.9.13.
jar:/software/hadoop-
2.6.5/share/hadoop/hdfs/lib/jersey-core-1.9.jar:/soft-
ware/hadoop-
2.6.5/share/hadoop/hdfs/lib/jetty-6.1.26.jar:/software/ha-
doop-
2.6.5/share/hadoop/hdfs/lib/log4j-1.2.17.jar:/software/ha-
doop-
2.6.5/share/hadoop/hdfs/lib/jsp-api-2.1.jar:/software/ha-
doop-
2.6.5/share/hadoop/hdfs/lib/xercesImpl-2.9.1.jar:/soft-
ware/hadoop-
2.6.5/share/hadoop/hdfs/hadoop-hdfs-2.6.5.jar:/software/
hadoop-
2.6.5/share/hadoop/hdfs/hadoop-hdfs-nfs-2.6.5.jar:/soft-
ware/hadoop-
2.6.5/share/hadoop/hdfs/hadoop-hdfs-2.6.5-tests.jar:/
software/hadoop-
2.6.5/share/hadoop/yarn/lib/commons-lang-2.6.jar:/soft-
ware/hadoop-
2.6.5/share/hadoop/yarn/lib/jackson-core-asl-1.9.13.
jar:/software/hadoop-
2.6.5/share/hadoop/yarn/lib/commons-codec-1.4.jar:/soft-
ware/hadoop-
2.6.5/share/hadoop/yarn/lib/jaxb-api-2.2.2.jar:/software/
hadoop-
2.6.5/share/hadoop/yarn/lib/javax.inject-1.jar:/software/
hadoop-
2.6.5/share/hadoop/yarn/lib/commons-cli-1.2.jar:/soft-
ware/hadoop-
```

```
2.6.5/share/hadoop/yarn/lib/jersey-guice-1.9.jar:/soft-
ware/hadoop-
2.6.5/share/hadoop/yarn/lib/jersey-json-1.9.jar:/soft-
ware/hadoop-
2.6.5/share/hadoop/yarn/lib/xz-1.0.jar:/software/hadoop-
2.6.5/share/hadoop/yarn/lib/zookeeper-3.4.6.jar:/software/
hadoop-
2.6.5/share/hadoop/yarn/lib/jersey-server-1.9.jar:/soft-
ware/hadoop-
2.6.5/share/hadoop/yarn/lib/commons-logging-1.1.3.jar:/
software/hadoop-
2.6.5/share/hadoop/yarn/lib/servlet-api-2.5.jar:/soft-
ware/hadoop-
2.6.5/share/hadoop/yarn/lib/netty-3.6.2.Final.jar:/soft-
ware/hadoop-
2.6.5/share/hadoop/yarn/lib/commons-compress-1.4.1.jar:/
software/hadoop-
2.6.5/share/hadoop/yarn/lib/protobuf-java-2.5.0.jar:/
software/hadoop-
2.6.5/share/hadoop/yarn/lib/leveldbjni-all-1.8.jar:/soft-
ware/hadoop-
2.6.5/share/hadoop/yarn/lib/activation-1.1.jar:/software/
hadoop-
2.6.5/share/hadoop/yarn/lib/jetty-util-6.1.26.jar:/soft-
ware/hadoop-
2.6.5/share/hadoop/yarn/lib/jackson-xc-1.9.13.jar:/soft-
ware/hadoop-
2.6.5/share/hadoop/yarn/lib/jsr305-1.3.9.jar:/software/ha-
doop-
2.6.5/share/hadoop/yarn/lib/jline-0.9.94.jar:/software/ha-
doop-
2.6.5/share/hadoop/yarn/lib/jaxb-impl-2.2.3-1.jar:/soft-
ware/hadoop-
2.6.5/share/hadoop/yarn/lib/guava-11.0.2.jar:/software/ha-
doop-
2.6.5/share/hadoop/yarn/lib/asm-3.2.jar:/software/hadoop-
2.6.5/share/hadoop/yarn/lib/commons-io-2.4.jar:/software/
hadoop-
```

```
2.6.5/share/hadoop/yarn/lib/jackson-mapper-asl-1.9.13.
jar:/software/hadoop-
2.6.5/share/hadoop/yarn/lib/commons-httpclient-3.1.jar:/
software/hadoop-
2.6.5/share/hadoop/yarn/lib/jersey-client-1.9.jar:/soft-
ware/hadoop-
2.6.5/share/hadoop/yarn/lib/guice-servlet-3.0.jar:/soft-
ware/hadoop-
2.6.5/share/hadoop/yarn/lib/jettison-1.1.jar:/software/ha-
doop-
2.6.5/share/hadoop/yarn/lib/guice-3.0.jar:/software/
hadoop-
2.6.5/share/hadoop/yarn/lib/stax-api-1.0-2.jar:/soft-
ware/hadoop-
2.6.5/share/hadoop/yarn/lib/jersey-core-1.9.jar:/soft-
ware/hadoop-
2.6.5/share/hadoop/yarn/lib/jackson-jaxrs-1.9.13.jar:/
software/hadoop-
2.6.5/share/hadoop/yarn/lib/jetty-6.1.26.jar:/software/ha-
doop-
2.6.5/share/hadoop/yarn/lib/log4j-1.2.17.jar:/software/ha-
doop-
2.6.5/share/hadoop/yarn/lib/aopalliance-1.0.jar:/software/
hadoop-
2.6.5/share/hadoop/yarn/lib/commons-collections-3.2.2.
jar:/software/hadoop-
2.6.5/share/hadoop/yarn/hadoop-yarn-server-web-proxy-
2.6.5.jar:/software/hadoop-
2.6.5/share/hadoop/yarn/hadoop-yarn-server-resourcemanager-
2.6.5.jar:/software/hadoop-2.6.5/share/hadoop/yarn/hadoop-
yarn-server-
applicationhistoryservice-2.6.5.jar:/software/hadoop-
2.6.5/share/hadoop/yarn/hadoop-yarn-client-2.6.5.jar:/
software/hadoop-2.6.5/share/hadoop/yarn/hadoop-yarn-reg-
istry-
2.6.5.jar:/software/hadoop-2.6.5/share/hadoop/yarn/hadoop-
yarn-common-
```

```
2.6.5.jar:/software/hadoop-2.6.5/share/hadoop/yarn/hadoop-
yarn-applications-
distributedshell-2.6.5.jar:/software/hadoop-2.6.5/share/
hadoop/yarn/hadoop-yarn-api-
2.6.5.jar:/software/hadoop-2.6.5/share/hadoop/yarn/hadoop-
yarn-server-common-
2.6.5.jar:/software/hadoop-2.6.5/share/hadoop/yarn/hadoop-
yarn-server-tests-
2.6.5.jar:/software/hadoop-2.6.5/share/hadoop/yarn/hadoop-
yarn-applications-
unmanaged-am-launcher-2.6.5.jar:/software/hadoop-2.6.5/
share/hadoop/yarn/hadoop-yarn-
server-nodemanager-2.6.5.jar:/software/hadoop-2.6.5/share/
hadoop/mapreduce/lib/avro-
1.7.4.jar:/software/hadoop-2.6.5/share/hadoop/mapreduce/lib/
jackson-core-asl-
1.9.13.jar:/software/hadoop-2.6.5/share/hadoop/mapreduce/
lib/javax.inject-
1.jar:/software/hadoop-2.6.5/share/hadoop/mapreduce/lib/
jersey-guice-
1.9.jar:/software/hadoop-2.6.5/share/hadoop/mapreduce/lib/xz-
1.0.jar:/software/hadoop-2.6.5/share/hadoop/mapreduce/lib/
paranamer-
2.3.jar:/software/hadoop-2.6.5/share/hadoop/mapreduce/lib/
jersey-server-
1.9.jar:/software/hadoop-2.6.5/share/hadoop/mapreduce/lib/
netty-
3.6.2.Final.jar:/software/hadoop-2.6.5/share/hadoop/mapre-
duce/lib/commons-compress-
1.4.1.jar:/software/hadoop-2.6.5/share/hadoop/mapreduce/
lib/protobuf-java-
2.5.0.jar:/software/hadoop-2.6.5/share/hadoop/mapreduce/
lib/snappy-java-
1.0.4.1.jar:/software/hadoop-2.6.5/share/hadoop/mapreduce/
lib/leveldbjni-all-
1.8.jar:/software/hadoop-2.6.5/share/hadoop/mapreduce/lib/
hamcrest-core-
```

```
1.3.jar:/software/hadoop-2.6.5/share/hadoop/mapreduce/lib/
hadoop-annotations-
2.6.5.jar:/software/hadoop-2.6.5/share/hadoop/mapreduce/
lib/asm-
3.2.jar:/software/hadoop-2.6.5/share/hadoop/mapreduce/lib/
commons-io-
2.4.jar:/software/hadoop-2.6.5/share/hadoop/mapreduce/lib/
junit-
4.11.jar:/software/hadoop-2.6.5/share/hadoop/mapreduce/
lib/jackson-mapper-asl-
1.9.13.jar:/software/hadoop-2.6.5/share/hadoop/mapreduce/
lib/guice-servlet-
3.0.jar:/software/hadoop-2.6.5/share/hadoop/mapreduce/lib/
guice-
3.0.jar:/software/hadoop-2.6.5/share/hadoop/mapreduce/lib/
jersey-core-
1.9.jar:/software/hadoop-2.6.5/share/hadoop/mapreduce/lib/
log4j-
1.2.17.jar:/software/hadoop-2.6.5/share/hadoop/mapreduce/
lib/aopalliance-
1.0.jar:/software/hadoop-2.6.5/share/hadoop/mapreduce/ha-
doop-mapreduce-examples-
2.6.5.jar:/software/hadoop-2.6.5/share/hadoop/mapreduce/
hadoop-mapreduce-client-
jobclient-2.6.5-tests.jar:/software/hadoop-2.6.5/share/
hadoop/mapreduce/hadoop-
mapreduce-client-shuffle-2.6.5.jar:/software/hadoop-
2.6.5/share/hadoop/mapreduce/hadoop-mapreduce-client-
core-
2.6.5.jar:/software/hadoop-2.6.5/share/hadoop/mapreduce/
hadoop-mapreduce-client-hs-plugins-
2.6.5.jar:/software/hadoop-2.6.5/share/hadoop/mapreduce/
hadoop-mapreduce-client-hs-
2.6.5.jar:/software/hadoop-2.6.5/share/hadoop/mapreduce/
hadoop-mapreduce-client-
common-2.6.5.jar:/software/hadoop-2.6.5/share/hadoop/ma-
preduce/hadoop-mapreduce-
```

```
client-app-2.6.5.jar:/software/hadoop-2.6.5/share/hadoop/
mapreduce/hadoop-mapreduce-
client-jobclient-2.6.5.jar:/software/hadoop-2.6.5/cont-
rib/capacity-
scheduler/*.jar19/04/14 22:11:14 INFO zookeeper.ZooKeeper: Client
environment:java.library.path=/software/hadoop-2.6.5/
lib/native
11.19/04/14 22:11:14 INFO zookeeper.ZooKeeper: Client environ-
ment:java.io.tmpdir=/tmp
12.19/04/14 22:11:14 INFO zookeeper.ZooKeeper: Client environ-
ment:java.compiler=<NA>
13.19/04/14 22:11:14 INFO zookeeper.ZooKeeper: Client environ-
ment:os.name=Linux
14.19/04/14 22:11:14 INFO zookeeper.ZooKeeper: Client environ-
ment:os.arch=amd64
15.19/04/14 22:11:14 INFO zookeeper.ZooKeeper: Client environ-
ment:os.version=2.6.32-504.el6.x86_64
16.19/04/14 22:11:14 INFO zookeeper.ZooKeeper: Client environ-
ment:user.name=root
17.19/04/14 22:11:14 INFO zookeeper.ZooKeeper: Client environ-
ment:user.home=/root
18.19/04/14 22:11:14 INFO zookeeper.ZooKeeper: Client environ-
ment:user.dir=/software
19.19/04/14 22:11:14 INFO zookeeper.ZooKeeper: Initiating client
connection, connectString=mynode3:2181,mynode4:2181,mynode5:
2181 sessionTimeout=5000 watcher=org.apache.hadoop.ha.Ac-
tiveStandbyElector$WatcherWithClientRef@6069db50
20.19/04/14 22:11:14 INFO zookeeper.ClientCnxn: Opening socket
connection to server mynode3/192.168.179.15:2181. Will not at-
tempt to authenticate using SASL (unknown error)
21.19/04/14 22:11:14 INFO zookeeper.ClientCnxn: Socket connection
established to mynode3/192.168.179.15:2181, initiating session
22.19/04/14 22:11:14 INFO zookeeper.ClientCnxn: Session estab-
lishment complete on server mynode3/192.168.179.15:2181, ses-
sionid=0x16a1c2ee2210000, negotiated timeout=5000
23.19/04/14 22:11:14 INFO ha.ActiveStandbyElector: Successfully
created /hadoop-ha/mycluster in ZK.
```

```
19/04/14 22:11:14 INFO zookeeper.ZooKeeper: Session: 0x16a1c
2ee2210000 closed
19/04/14 22:11:14 WARN ha.ActiveStandbyElector: Ignoring stale
result from old client with sessionId 0x16a1c2ee2210000
19/04/14 22:11:14 INFO zookeeper.ClientCnxn: EventThread shut
down
```

- 在每台 JournalNode 中启动所有的 JournalNode：

```
#在 mynode3 上启动 JournalNode
[root@mynode3 ~]# hadoop-daemon.sh start journalnode
starting journalnode, logging to /software/hadoop-2.6.5/logs/
hadoop-root-journalnode-mynode3.out

#在 mynode4 上启动 JournalNode
[root@mynode4 ~]# hadoop-daemon.sh start journalnode
starting journalnode, logging to /software/hadoop-2.6.5/logs/
hadoop-root-journalnode-mynode4.out

#在 mynode5 上启动 JournalNode
[root@mynode5 ~]# hadoop-daemon.sh start journalnode
starting journalnode, logging to /software/hadoop-2.6.5/
logs/hadoop-root-journalnode-mynode5.out
```

- 在 mynode1 上格式化 NameNode：

```
[root@mynode1 software]#hdfsnamenode -format
19/04/14 22:19:19 INFO namenode.NameNode: STARTUP_MSG:
/************************************************************
STARTUP_MSG: Starting NameNode
STARTUP_MSG: host = mynode1/192.168.179.13
STARTUP_MSG: args = [ -format]
STARTUP_MSG: version = 2.6.5
STARTUP_MSG: classpath = /software/hadoop-2.6.5/etc/hadoop:/
software/hadoop-2.6.5/share/hadoop/common/lib/avro-1.7.4.
jar:/software/hadoop-2.6.5/share/hadoop/common/lib/commons-
lang-
2.6.jar:/software/hadoop-2.6.5/share/hadoop/common/lib/
jackson-core-asl-
1.9.13.jar:/software/hadoop-2.6.5/share/hadoop/common/lib/
commons-codec-
```

```
1.4.jar:/software/hadoop-2.6.5/share/hadoop/common/lib/jaxb-
api-
2.2.2.jar:/software/hadoop-2.6.5/share/hadoop/common/lib/
commons-cli-
1.2.jar:/software/hadoop-2.6.5/share/hadoop/common/lib/com-
mons-configuration-
1.6.jar:/software/hadoop-2.6.5/share/hadoop/common/lib/
jersey-json-
1.9.jar:/software/hadoop-2.6.5/share/hadoop/common/lib/com-
mons-el-
1.0.jar:/software/hadoop-2.6.5/share/hadoop/common/lib/moc-
kito-all-
1.8.5.jar:/software/hadoop-2.6.5/share/hadoop/common/lib/
htrace-core-
3.0.4.jar:/software/hadoop-2.6.5/share/hadoop/common/lib/xz-
1.0.jar:/software/hadoop-
2.6.5/share/hadoop/common/lib/curator-framework-2.6.0.jar:/
software/hadoop-
2.6.5/share/hadoop/common/lib/zookeeper-3.4.6.jar:/soft-
ware/hadoop-
2.6.5/share/hadoop/common/lib/paranamer-2.3.jar:/software/
hadoop-
2.6.5/share/hadoop/common/lib/jersey-server-1.9.jar:/soft-
ware/hadoop-
2.6.5/share/hadoop/common/lib/commons-logging-1.1.3.jar:/
software/hadoop-
2.6.5/share/hadoop/common/lib/servlet-api-2.5.jar:/soft-
ware/hadoop-
2.6.5/share/hadoop/common/lib/apacheds-kerberos-codec-
2.0.0-M15.jar:/software/hadoop-
2.6.5/share/hadoop/common/lib/commons-net-3.1.jar:/soft-
ware/hadoop-
2.6.5/share/hadoop/common/lib/netty-3.6.2.Final.jar:/soft-
ware/hadoop-
2.6.5/share/hadoop/common/lib/commons-compress-1.4.1.jar:/
software/hadoop-
2.6.5/share/hadoop/common/lib/protobuf-java-2.5.0.jar:/
software/hadoop-
```

```
2.6.5/share/hadoop/common/lib/snappy-java-1.0.4.1.jar:/
software/hadoop-
2.6.5/share/hadoop/common/lib/api-util-1.0.0-M20.jar:/
software/hadoop-
2.6.5/share/hadoop/common/lib/commons-math3-3.1.1.jar:/
software/hadoop-
2.6.5/share/hadoop/common/lib/hamcrest-core-1.3.jar:/soft-
ware/hadoop-
2.6.5/share/hadoop/common/lib/commons-digester-1.8.jar:/
software/hadoop-
2.6.5/share/hadoop/common/lib/gson-2.2.4.jar:/software/ha-
doop-
2.6.5/share/hadoop/common/lib/activation-1.1.jar:/software/
hadoop-
2.6.5/share/hadoop/common/lib/xmlenc-0.52.jar:/software/ha-
doop-
2.6.5/share/hadoop/common/lib/jetty-util-6.1.26.jar:/soft-
ware/hadoop-
2.6.5/share/hadoop/common/lib/java-xmlbuilder-0.4.jar:/
software/hadoop-
2.6.5/share/hadoop/common/lib/hadoop-auth-2.6.5.jar:/soft-
ware/hadoop-
2.6.5/share/hadoop/common/lib/jackson-xc-1.9.13.jar:/soft-
ware/hadoop-
2.6.5/share/hadoop/common/lib/httpclient-4.2.5.jar:/soft-
ware/hadoop-
2.6.5/share/hadoop/common/lib/jsr305-1.3.9.jar:/software/
hadoop-
2.6.5/share/hadoop/common/lib/jasper-runtime-5.5.23.jar:/
software/hadoop-
2.6.5/share/hadoop/common/lib/commons-beanutils-1.7.0.jar:/
software/hadoop-
2.6.5/share/hadoop/common/lib/apacheds-i18n-2.0.0-M15.
jar:/software/hadoop-
2.6.5/share/hadoop/common/lib/jaxb-impl-2.2.3-1.jar:/soft-
ware/hadoop-
2.6.5/share/hadoop/common/lib/jsch-0.1.42.jar:/software/ha-
doop-
```

```
2.6.5/share/hadoop/common/lib/guava-11.0.2.jar:/software/
hadoop-
2.6.5/share/hadoop/common/lib/curator-client-2.6.0.jar:/
software/hadoop-
2.6.5/share/hadoop/common/lib/hadoop-annotations-2.6.5.
jar:/software/hadoop-
2.6.5/share/hadoop/common/lib/asm-3.2.jar:/software/
hadoop-
2.6.5/share/hadoop/common/lib/commons-io-2.4.jar:/soft-
ware/hadoop-
2.6.5/share/hadoop/common/lib/junit-4.11.jar:/software/ha-
doop-
2.6.5/share/hadoop/common/lib/jackson-mapper-asl-1.9.13.
jar:/software/hadoop-
2.6.5/share/hadoop/common/lib/curator-recipes-2.6.0.jar:/
software/hadoop-
2.6.5/share/hadoop/common/lib/commons-httpclient-3.1.jar:/
software/hadoop-
2.6.5/share/hadoop/common/lib/httpcore-4.2.5.jar:/software/
hadoop-
2.6.5/share/hadoop/common/lib/api-asn1-api-1.0.0-M20.
jar:/software/hadoop-
2.6.5/share/hadoop/common/lib/slf4j-api-1.7.5.jar:/soft-
ware/hadoop-
2.6.5/share/hadoop/common/lib/slf4j-log4j12-1.7.5.jar:/
software/hadoop-
2.6.5/share/hadoop/common/lib/jettison-1.1.jar:/software/
hadoop-
2.6.5/share/hadoop/common/lib/jasper-compiler-5.5.23.jar:/
software/hadoop-
2.6.5/share/hadoop/common/lib/stax-api-1.0-2.jar:/soft-
ware/hadoop-
2.6.5/share/hadoop/common/lib/jets3t-0.9.0.jar:/software/
hadoop-
2.6.5/share/hadoop/common/lib/jersey-core-1.9.jar:/soft-
ware/hadoop-
2.6.5/share/hadoop/common/lib/jackson-jaxrs-1.9.13.jar:/
software/hadoop-
```

```
2.6.5/share/hadoop/common/lib/commons - beanutils - core -
1.8.0.jar:/software/hadoop -
2.6.5/share/hadoop/common/lib/jetty -6.1.26.jar:/software/
hadoop -
2.6.5/share/hadoop/common/lib/log4j -1.2.17.jar:/software/
hadoop -
2.6.5/share/hadoop/common/lib/jsp - api -2.1.jar:/software/
hadoop -
2.6.5/share/hadoop/common/lib/commons - collections -3.2.2.
jar:/software/hadoop -
2.6.5/share/hadoop/common/hadoop - common -2.6.5 - tests.jar:/
software/hadoop -
2.6.5/share/hadoop/common/hadoop - common -2.6.5.jar:/soft-
ware/hadoop -
2.6.5/share/hadoop/common/hadoop - nfs -2.6.5.jar:/software/
hadoop -
2.6.5/share/hadoop/hdfs:/software/hadoop -2.6.5/share/ha-
doop/hdfs/lib/commons - lang -
2.6.jar:/software/hadoop -2.6.5/share/hadoop/hdfs/lib/jack-
son - core - asl -
1.9.13.jar:/software/hadoop -2.6.5/share/hadoop/hdfs/lib/
commons - codec -
1.4.jar:/software/hadoop -2.6.5/share/hadoop/hdfs/lib/com-
mons - cli -
1.2.jar:/software/hadoop -2.6.5/share/hadoop/hdfs/lib/xml -
apis -
1.3.04.jar:/software/hadoop -2.6.5/share/hadoop/hdfs/lib/
commons - el -
1.0.jar:/software/hadoop -2.6.5/share/hadoop/hdfs/lib/htrace
- core -
3.0.4.jar:/software/hadoop -2.6.5/share/hadoop/hdfs/lib/
jersey - server -
1.9.jar:/software/hadoop -2.6.5/share/hadoop/hdfs/lib/com-
mons - logging -
1.1.3.jar:/software/hadoop -2.6.5/share/hadoop/hdfs/lib/
servlet - api -
2.5.jar:/software/hadoop -2.6.5/share/hadoop/hdfs/lib/com-
mons - daemon -
```

```
1.0.13.jar:/software/hadoop-2.6.5/share/hadoop/hdfs/lib/
netty-
3.6.2.Final.jar:/software/hadoop-2.6.5/share/hadoop/hdfs/
lib/protobuf-java-
2.5.0.jar:/software/hadoop-2.6.5/share/hadoop/hdfs/lib/xm-
lenc-
0.52.jar:/software/hadoop-2.6.5/share/hadoop/hdfs/lib/jetty
-util-
6.1.26.jar:/software/hadoop-2.6.5/share/hadoop/hdfs/lib/
jsr305-
1.3.9.jar:/software/hadoop-2.6.5/share/hadoop/hdfs/lib/jas-
per-runtime-
5.5.23.jar:/software/hadoop-2.6.5/share/hadoop/hdfs/lib/
guava-
11.0.2.jar:/software/hadoop-2.6.5/share/hadoop/hdfs/lib/asm
-3.2.jar:/software/hadoop-
2.6.5/share/hadoop/hdfs/lib/commons-io-2.4.jar:/software/
hadoop-
2.6.5/share/hadoop/hdfs/lib/jackson-mapper-asl-1.9.13.
jar:/software/hadoop-
2.6.5/share/hadoop/hdfs/lib/jersey-core-1.9.jar:/software/
hadoop-
2.6.5/share/hadoop/hdfs/lib/jetty-6.1.26.jar:/software/ha-
doop-
2.6.5/share/hadoop/hdfs/lib/log4j-1.2.17.jar:/software/ha-
doop-
2.6.5/share/hadoop/hdfs/lib/jsp-api-2.1.jar:/software/ha-
doop-
2.6.5/share/hadoop/hdfs/lib/xercesImpl-2.9.1.jar:/software/
hadoop-
2.6.5/share/hadoop/hdfs/hadoop-hdfs-2.6.5.jar:/software/ha-
doop-
2.6.5/share/hadoop/hdfs/hadoop-hdfs-nfs-2.6.5.jar:/soft-
ware/hadoop-
2.6.5/share/hadoop/hdfs/hadoop-hdfs-2.6.5-tests.jar:/soft-
ware/hadoop-
2.6.5/share/hadoop/yarn/lib/commons-lang-2.6.jar:/soft-
ware/hadoop-
```

```
2.6.5/share/hadoop/yarn/lib/jackson-core-asl-1.9.13.jar:/
software/hadoop-
2.6.5/share/hadoop/yarn/lib/commons-codec-1.4.jar:/soft-
ware/hadoop-
2.6.5/share/hadoop/yarn/lib/jaxb-api-2.2.2.jar:/software/
hadoop-
2.6.5/share/hadoop/yarn/lib/javax.inject-1.jar:/software/
hadoop-
2.6.5/share/hadoop/yarn/lib/commons-cli-1.2.jar:/software/
hadoop-
2.6.5/share/hadoop/yarn/lib/jersey-guice-1.9.jar:/soft-
ware/hadoop-
2.6.5/share/hadoop/yarn/lib/jersey-json-1.9.jar:/software/
hadoop-
2.6.5/share/hadoop/yarn/lib/xz-1.0.jar:/software/hadoop-
2.6.5/share/hadoop/yarn/lib/zookeeper-3.4.6.jar:/software/
hadoop-
2.6.5/share/hadoop/yarn/lib/jersey-server-1.9.jar:/soft-
ware/hadoop-
2.6.5/share/hadoop/yarn/lib/commons-logging-1.1.3.jar:/
software/hadoop-
2.6.5/share/hadoop/yarn/lib/servlet-api-2.5.jar:/software/
hadoop-
2.6.5/share/hadoop/yarn/lib/netty-3.6.2.Final.jar:/soft-
ware/hadoop-
2.6.5/share/hadoop/yarn/lib/commons-compress-1.4.1.jar:/
software/hadoop-
2.6.5/share/hadoop/yarn/lib/protobuf-java-2.5.0.jar:/soft-
ware/hadoop-
2.6.5/share/hadoop/yarn/lib/leveldbjni-all-1.8.jar:/soft-
ware/hadoop-
2.6.5/share/hadoop/yarn/lib/activation-1.1.jar:/software/
hadoop-
2.6.5/share/hadoop/yarn/lib/jetty-util-6.1.26.jar:/soft-
ware/hadoop-
2.6.5/share/hadoop/yarn/lib/jackson-xc-1.9.13.jar:/soft-
ware/hadoop-
```

```
2.6.5/share/hadoop/yarn/lib/jsr305-1.3.9.jar:/software/ha-
doop-
2.6.5/share/hadoop/yarn/lib/jline-0.9.94.jar:/software/ha-
doop-
2.6.5/share/hadoop/yarn/lib/jaxb-impl-2.2.3-1.jar:/soft-
ware/hadoop-
2.6.5/share/hadoop/yarn/lib/guava-11.0.2.jar:/software/ha-
doop-
2.6.5/share/hadoop/yarn/lib/asm-3.2.jar:/software/hadoop-
2.6.5/share/hadoop/yarn/lib/commons-io-2.4.jar:/software/
hadoop-
2.6.5/share/hadoop/yarn/lib/jackson-mapper-asl-1.9.13.
jar:/software/hadoop-
2.6.5/share/hadoop/yarn/lib/commons-httpclient-3.1.jar:/
software/hadoop-
2.6.5/share/hadoop/yarn/lib/jersey-client-1.9.jar:/soft-
ware/hadoop-
2.6.5/share/hadoop/yarn/lib/guice-servlet-3.0.jar:/soft-
ware/hadoop-
2.6.5/share/hadoop/yarn/lib/jettison-1.1.jar:/software/ha-
doop-
2.6.5/share/hadoop/yarn/lib/guice-3.0.jar:/software/
hadoop-
2.6.5/share/hadoop/yarn/lib/stax-api-1.0-2.jar:/software/
hadoop-
2.6.5/share/hadoop/yarn/lib/jersey-core-1.9.jar:/software/
hadoop-
2.6.5/share/hadoop/yarn/lib/jackson-jaxrs-1.9.13.jar:/soft-
ware/hadoop-
2.6.5/share/hadoop/yarn/lib/jetty-6.1.26.jar:/software/ha-
doop-
2.6.5/share/hadoop/yarn/lib/log4j-1.2.17.jar:/software/ha-
doop-
2.6.5/share/hadoop/yarn/lib/aopalliance-1.0.jar:/software/
hadoop-
2.6.5/share/hadoop/yarn/lib/commons-collections-3.2.2.jar:/
software/hadoop-
```

```
2.6.5/share/hadoop/yarn/hadoop-yarn-server-web-proxy-
2.6.5.jar:/software/hadoop-
2.6.5/share/hadoop/yarn/hadoop-yarn-server-resourcemanager-
2.6.5.jar:/software/hadoop-2.6.5/share/hadoop/yarn/hadoop-
yarn-server-applicationhistoryservice-
2.6.5.jar:/software/hadoop-2.6.5/share/hadoop/yarn/hadoop-
yarn-client-
2.6.5.jar:/software/hadoop-2.6.5/share/hadoop/yarn/hadoop-
yarn-registry-
2.6.5.jar:/software/hadoop-2.6.5/share/hadoop/yarn/hadoop-
yarn-common-
2.6.5.jar:/software/hadoop-2.6.5/share/hadoop/yarn/hadoop-
yarn-applications-distributedshell-
2.6.5.jar:/software/hadoop-2.6.5/share/hadoop/yarn/hadoop-
yarn-api-
2.6.5.jar:/software/hadoop-2.6.5/share/hadoop/yarn/hadoop-
yarn-server-common-
2.6.5.jar:/software/hadoop-2.6.5/share/hadoop/yarn/hadoop-
yarn-server-tests-
2.6.5.jar:/software/hadoop-2.6.5/share/hadoop/yarn/hadoop-
yarn-applications-unmanaged-am-launcher-
2.6.5.jar:/software/hadoop-2.6.5/share/hadoop/yarn/hadoop-
yarn-server-nodemanager-
2.6.5.jar:/software/hadoop-2.6.5/share/hadoop/mapreduce/
lib/avro-
1.7.4.jar:/software/hadoop-2.6.5/share/hadoop/mapreduce/
lib/jackson-core-asl-
1.9.13.jar:/software/hadoop-2.6.5/share/hadoop/mapreduce/
lib/javax.inject-
1.jar:/software/hadoop-2.6.5/share/hadoop/mapreduce/lib/
jersey-guice-
1.9.jar:/software/hadoop-2.6.5/share/hadoop/mapreduce/lib/
xz-
1.0.jar:/software/hadoop-2.6.5/share/hadoop/mapreduce/lib/
paranamer-
2.3.jar:/software/hadoop-2.6.5/share/hadoop/mapreduce/lib/
jersey-server-
```

```
1.9.jar:/software/hadoop-2.6.5/share/hadoop/mapreduce/lib/
netty-
3.6.2.Final.jar:/software/hadoop-2.6.5/share/hadoop/mapre-
duce/lib/commons-compress-
1.4.1.jar:/software/hadoop-2.6.5/share/hadoop/mapreduce/
lib/protobuf-java-
2.5.0.jar:/software/hadoop-2.6.5/share/hadoop/mapreduce/
lib/snappy-java-
1.0.4.1.jar:/software/hadoop-2.6.5/share/hadoop/mapreduce/
lib/leveldbjni-all-
1.8.jar:/software/hadoop-2.6.5/share/hadoop/mapreduce/lib/
hamcrest-core-
1.3.jar:/software/hadoop-2.6.5/share/hadoop/mapreduce/lib/
hadoop-annotations-
2.6.5.jar:/software/hadoop-2.6.5/share/hadoop/mapreduce/
lib/asm-
3.2.jar:/software/hadoop-2.6.5/share/hadoop/mapreduce/lib/
commons-io-
2.4.jar:/software/hadoop-2.6.5/share/hadoop/mapreduce/lib/
junit-
4.11.jar:/software/hadoop-2.6.5/share/hadoop/mapreduce/lib/
jackson-mapper-asl-
1.9.13.jar:/software/hadoop-2.6.5/share/hadoop/mapreduce/
lib/guice-servlet-
3.0.jar:/software/hadoop-2.6.5/share/hadoop/mapreduce/lib/
guice-
3.0.jar:/software/hadoop-2.6.5/share/hadoop/mapreduce/lib/
jersey-core-
1.9.jar:/software/hadoop-2.6.5/share/hadoop/mapreduce/lib/
log4j-
1.2.17.jar:/software/hadoop-2.6.5/share/hadoop/mapreduce/
lib/aopalliance-
1.0.jar:/software/hadoop-2.6.5/share/hadoop/mapreduce/ha-
doop-mapreduce-examples-
2.6.5.jar:/software/hadoop-2.6.5/share/hadoop/mapreduce/ha-
doop-mapreduce-client-jobclient-
2.6.5-tests.jar:/software/hadoop-2.6.5/share/hadoop/mapre-
duce/hadoop-
```

```
    mapreduce-client-shuffle-2.6.5.jar:/software/hadoop-
    2.6.5/share/hadoop/mapreduce/hadoop-mapreduce-client-core-
    2.6.5.jar:/software/hadoop-
    2.6.5/share/hadoop/mapreduce/hadoop-mapreduce-client-hs-
    plugins-2.6.5.jar:/software/hadoop-
    2.6.5/share/hadoop/mapreduce/hadoop-mapreduce-client-hs-
    2.6.5.jar:/software/hadoop-
    2.6.5/share/hadoop/mapreduce/hadoop-mapreduce-client-
    common-2.6.5.jar:/software/hadoop-
    2.6.5/share/hadoop/mapreduce/hadoop-mapreduce-client-app-
    2.6.5.jar:/software/hadoop-
    2.6.5/share/hadoop/mapreduce/hadoop-mapreduce-client-job-
    client-
    2.6.5.jar:/software/hadoop-
    2.6.5/contrib/capacity-
    scheduler/*.jarSTARTUP_MSG: build=https://github.com/apache/
    hadoop.git -r
    e8c9fe0b4c252caf2ebf1464220599650f119997; compiled by 'sjlee'
    on 2016-10-02T23:43Z
10. STARTUP_MSG: java=1.8.0_181
11. ************************************************************/
12. 19/04/14 22:19:19 INFO namenode.NameNode: registered UNIX sig-
    nal handlers for [TERM, HUP, INT]
13. 19/04/14 22:19:19 INFO namenode.NameNode: createNameNode [-fo-
    rmat]
14. 19/04/14 22:19:20 WARN util.NativeCodeLoader: Unable to load
    native-hadoop library for your platform... using builtin-java
    classes where applicable
15. Formatting using clusterid: CID-90ae45f6-03bd-4c69-8b37-
    d04b07718e7d
16. 19/04/14 22:19:20 INFO namenode.FSNamesystem: No KeyProvider
    found.
17. 19/04/14 22:19:20 INFO namenode.FSNamesystem: fsLock is
    fair:true
18. 19/04/14 22:19:20 INFO blockmanagement.DatanodeManager: dfs.
    block.invalidate.limit=1000
19. 19/04/14 22:19:20 INFO blockmanagement.DatanodeManager: dfs.
    namenode.datanode.registration.ip-hostname-check=true
```

```
19/04/14 22:19:20 INFO blockmanagement.BlockManager: dfs.namenode.startup.delay.block.deletion.sec is set to 000:00:00:00.000
19/04/14 22:19:20 INFO blockmanagement.BlockManager: The block deletion will start around 2019 四月 14 22:19:20
19/04/14 22:19:20 INFO util.GSet: Computing capacity for map BlocksMap
19/04/14 22:19:20 INFO util.GSet: VM type = 64-bit
19/04/14 22:19:20 INFO util.GSet: 2.0% max memory 889 MB = 17.8 MB
19/04/14 22:19:20 INFO util.GSet: capacity = 2^21 = 2097152 entries
19/04/14 22:19:20 INFO blockmanagement.BlockManager: dfs.block.access.token.enable = false
19/04/14 22:19:20 INFO blockmanagement.BlockManager: defaultReplication = 3
19/04/14 22:19:20 INFO blockmanagement.BlockManager: maxReplication = 512
19/04/14 22:19:20 INFO blockmanagement.BlockManager: minReplication = 1
19/04/14 22:19:20 INFO blockmanagement.BlockManager: maxReplicationStreams = 2
19/04/14 22:19:20 INFO blockmanagement.BlockManager: replicationRecheckInterval = 3000
19/04/14 22:19:20 INFO blockmanagement.BlockManager: encryptDataTransfer = false
19/04/14 22:19:20 INFO blockmanagement.BlockManager: maxNumBlocksToLog = 1000
19/04/14 22:19:20 INFO namenode.FSNamesystem: fsOwner = root (auth:SIMPLE)
19/04/14 22:19:20 INFO namenode.FSNamesystem: supergroup = supergroup
19/04/14 22:19:20 INFO namenode.FSNamesystem: isPermissionEnabled = false
19/04/14 22:19:20 INFO namenode.FSNamesystem: Determined nameservice ID: mycluster
19/04/14 22:19:20 INFO namenode.FSNamesystem: HA Enabled: true
19/04/14 22:19:20 INFO namenode.FSNamesystem: Append Enabled: true
```

```
40. 19/04/14 22:19:20 INFO util.GSet: Computing capacity for map IN-
    odeMap
41. 19/04/14 22:19:20 INFO util.GSet: VM type = 64-bit
42. 19/04/14 22:19:20 INFO util.GSet: 1.0% max memory 889 MB = 8.9 MB
43. 19/04/14 22:19:20 INFO util.GSet: capacity = 2^20 = 1048576 en-
    tries
44. 19/04/14 22:19:20 INFO namenode.NameNode: Caching file names
    occuring more than 10 times
45. 19/04/14 22:19:20 INFO util.GSet: Computing capacity for
    map cachedBlocks
46. 19/04/14 22:19:20 INFO util.GSet: VM type = 64-bit
47. 19/04/14 22:19:20 INFO util.GSet: 0.25% max memory 889 MB = 2.2 MB
48. 19/04/14 22:19:20 INFO util.GSet: capacity = 2^18 = 262144 en-
    tries
49. 19/04/14 22:19:20 INFO namenode.FSNamesystem: dfs.namenode.
    safemode.threshold-pct = 0.9990000128746033
50. 19/04/14 22:19:20 INFO namenode.FSNamesystem: dfs.namenode.
    safemode.min.datanodes = 0
51. 19/04/14 22:19:20 INFO namenode.FSNamesystem: dfs.namenode.
    safemode.extension = 30000
52. 19/04/14 22:19:20 INFO namenode.FSNamesystem: Retry cache on
    namenode is enabled
53. 19/04/14 22:19:20 INFO namenode.FSNamesystem: Retry cache will
    use 0.03 of total heap and retry cache entry expiry time is
    600000 millis
54. 19/04/14 22:19:20 INFO util.GSet: Computing capacity for map Na-
    meNodeRetryCache
55. 19/04/14 22:19:20 INFO util.GSet: VM type = 64-bit
56. 19/04/14 22:19:20 INFO util.GSet: 0.029999999329447746% max
    memory 889 MB = 273.1 KB
57. 19/04/14 22:19:20 INFO util.GSet: capacity = 2^15 = 32768 entries
58. 19/04/14 22:19:20 INFO namenode.NNConf: ACLs enabled? false
59. 19/04/14 22:19:20 INFO namenode.NNConf: XAttrs enabled? true
60. 19/04/14 22:19:20 INFO namenode.NNConf: Maximum size of an xat-
    tr: 16384
61. 19/04/14 22:19:21 INFO namenode.FSImage: Allocated new Block-
    PoolId: BP-481747682-192.168.179.13-1555251561577
```

```
19/04/14 22:19:21 INFO common.Storage: Storage directory /opt/data/hadoop/dfs/name has been successfully formatted.
19/04/14 22:19:21 INFO namenode.FSImageFormatProtobuf: Saving image file /opt/data/hadoop/dfs/name/current/fsimage.ckpt_0000000000000000000 using no compression
19/04/14 22:19:21 INFO namenode.FSImageFormatProtobuf: Image file /opt/data/hadoop/dfs/name/current/fsimage.ckpt_0000000000000000000 of size 321 bytes saved in 0 seconds.
19/04/14 22:19:21 INFO namenode.NNStorageRetentionManager: Going to retain 1 images with txid >= 0
19/04/14 22:19:21 INFO util.ExitUtil: Exiting with status 0
19/04/14 22:19:21 INFO namenode.NameNode: SHUTDOWN_MSG:
/************************************************************
SHUTDOWN_MSG: Shutting down NameNode at mynode1/192.168.179.13
************************************************************/
```

- 在 mynode1 启动 NameNode，同时，在 mynode2 中执行同步：

```
#在 mynode1 中启动 NameNode
[root@mynode1 software]# hadoop-daemon.sh start namenode
starting namenode, logging to /software/hadoop-2.6.5/logs/hadoop-root-namenode-mynode1.out

#在 mynode2 中执行同步:
[root@mynode2 ~]#hdfsnamenode -bootstrapStandby
19/04/14 22:24:01 INFO namenode.NameNode: STARTUP_MSG:
/************************************************************
STARTUP_MSG: Starting NameNode
STARTUP_MSG: host = mynode2/192.168.179.14
STARTUP_MSG: args = [-bootstrapStandby]
STARTUP_MSG: version = 2.6.5
STARTUP_MSG: classpath = /software/hadoop-2.6.5/etc/hadoop:/software/hadoop-
2.6.5/share/hadoop/common/lib/avro-1.7.4.jar:/software/hadoop-
2.6.5/share/hadoop/common/lib/commons-lang-2.6.jar:/software/hadoop-
2.6.5/share/hadoop/common/lib/jackson-core-asl-
```

```
1.9.13.jar:/software/hadoop-2.6.5/share/hadoop/common/lib/
commons-codec-
1.4.jar:/software/hadoop-2.6.5/share/hadoop/common/lib/
jaxb-api-
2.2.2.jar:/software/hadoop-2.6.5/share/hadoop/common/lib/
commons-cli-
1.2.jar:/software/hadoop-2.6.5/share/hadoop/common/lib/
commons-configuration-
1.6.jar:/software/hadoop-2.6.5/share/hadoop/common/lib/
jersey-json-
1.9.jar:/software/hadoop-2.6.5/share/hadoop/common/lib/
commons-el-
1.0.jar:/software/hadoop-2.6.5/share/hadoop/common/lib/
mockito-all-
1.8.5.jar:/software/hadoop-2.6.5/share/hadoop/common/lib/
htrace-core-
3.0.4.jar:/software/hadoop-2.6.5/share/hadoop/common/lib/
xz-1.0.jar:/software/hadoop-
2.6.5/share/hadoop/common/lib/curator-framework-2.6.0.
jar:/software/hadoop-
2.6.5/share/hadoop/common/lib/zookeeper-3.4.6.jar:/soft-
ware/hadoop-
2.6.5/share/hadoop/common/lib/paranamer-2.3.jar:/software/
hadoop-
2.6.5/share/hadoop/common/lib/jersey-server-1.9.jar:/soft-
ware/hadoop-
2.6.5/share/hadoop/common/lib/commons-logging-1.1.3.jar:/
software/hadoop-
2.6.5/share/hadoop/common/lib/servlet-api-2.5.jar:/soft-
ware/hadoop-
2.6.5/share/hadoop/common/lib/apacheds-kerberos-codec-
2.0.0-M15.jar:/software/hadoop-
2.6.5/share/hadoop/common/lib/commons-net-3.1.jar:/soft-
ware/hadoop-
2.6.5/share/hadoop/common/lib/netty-3.6.2.Final.jar:/soft-
ware/hadoop-
2.6.5/share/hadoop/common/lib/commons-compress-1.4.1.jar:/
software/hadoop-
```

```
2.6.5/share/hadoop/common/lib/protobuf-java-2.5.0.jar:/
software/hadoop-
2.6.5/share/hadoop/common/lib/snappy-java-1.0.4.1.jar:/
software/hadoop-
2.6.5/share/hadoop/common/lib/api-util-1.0.0-M20.jar:/
software/hadoop-
2.6.5/share/hadoop/common/lib/commons-math3-3.1.1.jar:/
software/hadoop-
2.6.5/share/hadoop/common/lib/hamcrest-core-1.3.jar:/soft-
ware/hadoop-
2.6.5/share/hadoop/common/lib/commons-digester-1.8.jar:/
software/hadoop-
2.6.5/share/hadoop/common/lib/gson-2.2.4.jar:/software/
hadoop-
2.6.5/share/hadoop/common/lib/activation-1.1.jar:/soft-
ware/hadoop-
2.6.5/share/hadoop/common/lib/xmlenc-0.52.jar:/software/
hadoop-
2.6.5/share/hadoop/common/lib/jetty-util-6.1.26.jar:/soft-
ware/hadoop-
2.6.5/share/hadoop/common/lib/java-xmlbuilder-0.4.jar:/
software/hadoop-
2.6.5/share/hadoop/common/lib/hadoop-auth-2.6.5.jar:/soft-
ware/hadoop-
2.6.5/share/hadoop/common/lib/jackson-xc-1.9.13.jar:/soft-
ware/hadoop-
2.6.5/share/hadoop/common/lib/httpclient-4.2.5.jar:/soft-
ware/hadoop-
2.6.5/share/hadoop/common/lib/jsr305-1.3.9.jar:/software/
hadoop-
2.6.5/share/hadoop/common/lib/jasper-runtime-5.5.23.jar:/
software/hadoop-
2.6.5/share/hadoop/common/lib/commons-beanutils-1.7.0.
jar:/software/hadoop-
2.6.5/share/hadoop/common/lib/apacheds-i18n-2.0.0-M15.
jar:/software/hadoop-
2.6.5/share/hadoop/common/lib/jaxb-impl-2.2.3-1.jar:/
software/hadoop-
```

```
2.6.5/share/hadoop/common/lib/jsch-0.1.42.jar:/software/
hadoop-
2.6.5/share/hadoop/common/lib/guava-11.0.2.jar:/software/
hadoop-
2.6.5/share/hadoop/common/lib/curator-client-2.6.0.jar:/
software/hadoop-
2.6.5/share/hadoop/common/lib/hadoop-annotations-2.6.5.
jar:/software/hadoop-
2.6.5/share/hadoop/common/lib/asm-3.2.jar:/software/hadoop-
2.6.5/share/hadoop/common/lib/commons-io-2.4.jar:/soft-
ware/hadoop-
2.6.5/share/hadoop/common/lib/junit-4.11.jar:/software/ha-
doop-
2.6.5/share/hadoop/common/lib/jackson-mapper-asl-1.9.13.
jar:/software/hadoop-
2.6.5/share/hadoop/common/lib/curator-recipes-2.6.0.jar:/
software/hadoop-
2.6.5/share/hadoop/common/lib/commons-httpclient-3.1.jar:/
software/hadoop-
2.6.5/share/hadoop/common/lib/httpcore-4.2.5.jar:/soft-
ware/hadoop-
2.6.5/share/hadoop/common/lib/api-asn1-api-1.0.0-M20.
jar:/software/hadoop-
2.6.5/share/hadoop/common/lib/slf4j-api-1.7.5.jar:/soft-
ware/hadoop-
2.6.5/share/hadoop/common/lib/slf4j-log4j12-1.7.5.jar:/
software/hadoop-
2.6.5/share/hadoop/common/lib/jettison-1.1.jar:/software/
hadoop-
2.6.5/share/hadoop/common/lib/jasper-compiler-5.5.23.jar:/
software/hadoop-
2.6.5/share/hadoop/common/lib/stax-api-1.0-2.jar:/soft-
ware/hadoop-
2.6.5/share/hadoop/common/lib/jets3t-0.9.0.jar:/software/
hadoop-
2.6.5/share/hadoop/common/lib/jersey-core-1.9.jar:/soft-
ware/hadoop-
```

```
2.6.5/share/hadoop/common/lib/jackson-jaxrs-1.9.13.jar:/
software/hadoop-
2.6.5/share/hadoop/common/lib/commons-beanutils-core-1.8.0.
jar:/software/hadoop-
2.6.5/share/hadoop/common/lib/jetty-6.1.26.jar:/software/
hadoop-
2.6.5/share/hadoop/common/lib/log4j-1.2.17.jar:/software/
hadoop-
2.6.5/share/hadoop/common/lib/jsp-api-2.1.jar:/software/
hadoop-
2.6.5/share/hadoop/common/lib/commons-collections-3.2.2.
jar:/software/hadoop-
2.6.5/share/hadoop/common/hadoop-common-2.6.5-tests.jar:/
software/hadoop-
2.6.5/share/hadoop/common/hadoop-common-2.6.5.jar:/soft-
ware/hadoop-
2.6.5/share/hadoop/common/hadoop-nfs-2.6.5.jar:/software/
hadoop-
2.6.5/share/hadoop/hdfs:/software/hadoop-2.6.5/share/
hadoop/hdfs/lib/commons-lang-
2.6.jar:/software/hadoop-2.6.5/share/hadoop/hdfs/lib/jack-
son-core-asl-
1.9.13.jar:/software/hadoop-2.6.5/share/hadoop/hdfs/lib/
commons-codec-
1.4.jar:/software/hadoop-2.6.5/share/hadoop/hdfs/lib/com-
mons-cli-
1.2.jar:/software/hadoop-2.6.5/share/hadoop/hdfs/lib/xml-
apis-
1.3.04.jar:/software/hadoop-2.6.5/share/hadoop/hdfs/lib/
commons-el-
1.0.jar:/software/hadoop-2.6.5/share/hadoop/hdfs/lib/
htrace-core-
3.0.4.jar:/software/hadoop-2.6.5/share/hadoop/hdfs/lib/
jersey-server-
1.9.jar:/software/hadoop-2.6.5/share/hadoop/hdfs/lib/com-
mons-logging-
1.1.3.jar:/software/hadoop-2.6.5/share/hadoop/hdfs/lib/
servlet-api-
```

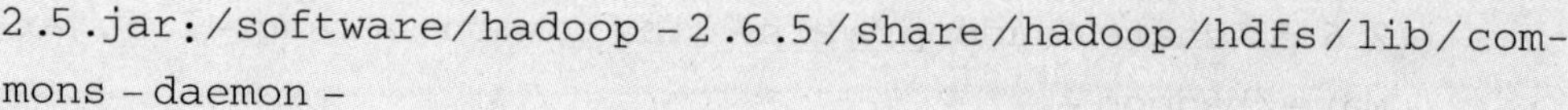



```
2.5.jar:/software/hadoop-2.6.5/share/hadoop/hdfs/lib/com-
mons-daemon-
1.0.13.jar:/software/hadoop-2.6.5/share/hadoop/hdfs/lib/
netty-
3.6.2.Final.jar:/software/hadoop-2.6.5/share/hadoop/hdfs/
lib/protobuf-java-
2.5.0.jar:/software/hadoop-2.6.5/share/hadoop/hdfs/lib/
xmlenc-
0.52.jar:/software/hadoop-2.6.5/share/hadoop/hdfs/lib/
jetty-util-
6.1.26.jar:/software/hadoop-2.6.5/share/hadoop/hdfs/lib/
jsr305-
1.3.9.jar:/software/hadoop-2.6.5/share/hadoop/hdfs/lib/
jasper-runtime-
5.5.23.jar:/software/hadoop-2.6.5/share/hadoop/hdfs/lib/
guava-
11.0.2.jar:/software/hadoop-2.6.5/share/hadoop/hdfs/lib/
asm-3.2.jar:/software/hadoop-
2.6.5/share/hadoop/hdfs/lib/commons-io-2.4.jar:/software/
hadoop-
2.6.5/share/hadoop/hdfs/lib/jackson-mapper-asl-1.9.13.
jar:/software/hadoop-
2.6.5/share/hadoop/hdfs/lib/jersey-core-1.9.jar:/soft-
ware/hadoop-
2.6.5/share/hadoop/hdfs/lib/jetty-6.1.26.jar:/software/
hadoop-
2.6.5/share/hadoop/hdfs/lib/log4j-1.2.17.jar:/software/
hadoop-
2.6.5/share/hadoop/hdfs/lib/jsp-api-2.1.jar:/software/
hadoop-
2.6.5/share/hadoop/hdfs/lib/xercesImpl-2.9.1.jar:/soft-
ware/hadoop-
2.6.5/share/hadoop/hdfs/hadoop-hdfs-2.6.5.jar:/software/
hadoop-
2.6.5/share/hadoop/hdfs/hadoop-hdfs-nfs-2.6.5.jar:/soft-
ware/hadoop-
2.6.5/share/hadoop/hdfs/hadoop-hdfs-2.6.5-tests.jar:/
software/hadoop-
```

```
2.6.5/share/hadoop/yarn/lib/commons-lang-2.6.jar:/soft-
ware/hadoop-
2.6.5/share/hadoop/yarn/lib/jackson-core-asl-1.9.13.jar:/
software/hadoop-
2.6.5/share/hadoop/yarn/lib/commons-codec-1.4.jar:/soft-
ware/hadoop-
2.6.5/share/hadoop/yarn/lib/jaxb-api-2.2.2.jar:/software/
hadoop-
2.6.5/share/hadoop/yarn/lib/javax.inject-1.jar:/software/
hadoop-
2.6.5/share/hadoop/yarn/lib/commons-cli-1.2.jar:/soft-
ware/hadoop-
2.6.5/share/hadoop/yarn/lib/jersey-guice-1.9.jar:/soft-
ware/hadoop-
2.6.5/share/hadoop/yarn/lib/jersey-json-1.9.jar:/soft-
ware/hadoop-
2.6.5/share/hadoop/yarn/lib/xz-1.0.jar:/software/hadoop-
2.6.5/share/hadoop/yarn/lib/zookeeper-3.4.6.jar:/software/
hadoop-
2.6.5/share/hadoop/yarn/lib/jersey-server-1.9.jar:/soft-
ware/hadoop-
2.6.5/share/hadoop/yarn/lib/commons-logging-1.1.3.jar:/
software/hadoop-
2.6.5/share/hadoop/yarn/lib/servlet-api-2.5.jar:/soft-
ware/hadoop-
2.6.5/share/hadoop/yarn/lib/netty-3.6.2.Final.jar:/soft-
ware/hadoop-
2.6.5/share/hadoop/yarn/lib/commons-compress-1.4.1.jar:/
software/hadoop-
2.6.5/share/hadoop/yarn/lib/protobuf-java-2.5.0.jar:/soft-
ware/hadoop-
2.6.5/share/hadoop/yarn/lib/leveldbjni-all-1.8.jar:/soft-
ware/hadoop-
2.6.5/share/hadoop/yarn/lib/activation-1.1.jar:/software/
hadoop-
2.6.5/share/hadoop/yarn/lib/jetty-util-6.1.26.jar:/soft-
ware/hadoop-
```

```
2.6.5/share/hadoop/yarn/lib/jackson-xc-1.9.13.jar:/soft-
ware/hadoop-
2.6.5/share/hadoop/yarn/lib/jsr305-1.3.9.jar:/software/
hadoop-
2.6.5/share/hadoop/yarn/lib/jline-0.9.94.jar:/software/
hadoop-
2.6.5/share/hadoop/yarn/lib/jaxb-impl-2.2.3-1.jar:/soft-
ware/hadoop-
2.6.5/share/hadoop/yarn/lib/guava-11.0.2.jar:/software/
hadoop-
2.6.5/share/hadoop/yarn/lib/asm-3.2.jar:/software/hadoop-
2.6.5/share/hadoop/yarn/lib/commons-io-2.4.jar:/software/
hadoop-
2.6.5/share/hadoop/yarn/lib/jackson-mapper-asl-1.9.13.
jar:/software/hadoop-
2.6.5/share/hadoop/yarn/lib/commons-httpclient-3.1.jar:/
software/hadoop-
2.6.5/share/hadoop/yarn/lib/jersey-client-1.9.jar:/soft-
ware/hadoop-
2.6.5/share/hadoop/yarn/lib/guice-servlet-3.0.jar:/soft-
ware/hadoop-
2.6.5/share/hadoop/yarn/lib/jettison-1.1.jar:/software/
hadoop-
2.6.5/share/hadoop/yarn/lib/guice-3.0.jar:/software/hadoop-
2.6.5/share/hadoop/yarn/lib/stax-api-1.0-2.jar:/software/
hadoop-
2.6.5/share/hadoop/yarn/lib/jersey-core-1.9.jar:/soft-
ware/hadoop-
2.6.5/share/hadoop/yarn/lib/jackson-jaxrs-1.9.13.jar:/
software/hadoop-
2.6.5/share/hadoop/yarn/lib/jetty-6.1.26.jar:/software/
hadoop-
2.6.5/share/hadoop/yarn/lib/log4j-1.2.17.jar:/software/
hadoop-
2.6.5/share/hadoop/yarn/lib/aopalliance-1.0.jar:/software/
hadoop-
2.6.5/share/hadoop/yarn/lib/commons-collections-3.2.2.
jar:/software/hadoop-
```

```
2.6.5/share/hadoop/yarn/hadoop-yarn-server-web-proxy-
2.6.5.jar:/software/hadoop-
2.6.5/share/hadoop/yarn/hadoop-yarn-server-resourcemanager-
2.6.5.jar:/software/hadoop-2.6.5/share/hadoop/yarn/hadoop-
yarn-server-
applicationhistoryservice-2.6.5.jar:/software/hadoop-2.6.5/
share/hadoop/yarn/hadoop-
yarn-client-2.6.5.jar:/software/hadoop-2.6.5/share/ha-
doop/yarn/hadoop-yarn-registry-
2.6.5.jar:/software/hadoop-2.6.5/share/hadoop/yarn/hadoop-
yarn-common-
2.6.5.jar:/software/hadoop-2.6.5/share/hadoop/yarn/hadoop-
yarn-applications-
distributedshell-2.6.5.jar:/software/hadoop-2.6.5/share/
hadoop/yarn/hadoop-yarn-api-
2.6.5.jar:/software/hadoop-2.6.5/share/hadoop/yarn/hadoop-
yarn-server-common-
2.6.5.jar:/software/hadoop-2.6.5/share/hadoop/yarn/hadoop-
yarn-server-tests-
2.6.5.jar:/software/hadoop-2.6.5/share/hadoop/yarn/hadoop-
yarn-applications-unmanaged-am-launcher-
2.6.5.jar:/software/hadoop-2.6.5/share/hadoop/yarn/hadoop-
yarn-server-nodemanager-
2.6.5.jar:/software/hadoop-2.6.5/share/hadoop/mapreduce/
lib/avro-
1.7.4.jar:/software/hadoop-2.6.5/share/hadoop/mapreduce/
lib/jackson-core-asl-
1.9.13.jar:/software/hadoop-2.6.5/share/hadoop/mapreduce/
lib/javax.inject-
1.jar:/software/hadoop-2.6.5/share/hadoop/mapreduce/lib/
jersey-guice-
1.9.jar:/software/hadoop-2.6.5/share/hadoop/mapreduce/lib/
xz-
1.0.jar:/software/hadoop-2.6.5/share/hadoop/mapreduce/lib/
paranamer-
2.3.jar:/software/hadoop-2.6.5/share/hadoop/mapreduce/lib/
jersey-server-
```

```
1.9.jar:/software/hadoop-2.6.5/share/hadoop/mapreduce/lib/
netty-
3.6.2.Final.jar:/software/hadoop-2.6.5/share/hadoop/mapre-
duce/lib/commons-compress-
1.4.1.jar:/software/hadoop-2.6.5/share/hadoop/mapreduce/
lib/protobuf-java-
2.5.0.jar:/software/hadoop-2.6.5/share/hadoop/mapreduce/
lib/snappy-java-
1.0.4.1.jar:/software/hadoop-2.6.5/share/hadoop/mapreduce/
lib/leveldbjni-all-
1.8.jar:/software/hadoop-2.6.5/share/hadoop/mapreduce/lib/
hamcrest-core-
1.3.jar:/software/hadoop-2.6.5/share/hadoop/mapreduce/lib/
hadoop-annotations-
2.6.5.jar:/software/hadoop-2.6.5/share/hadoop/mapreduce/
lib/asm-
3.2.jar:/software/hadoop-2.6.5/share/hadoop/mapreduce/lib/
commons-io-
2.4.jar:/software/hadoop-2.6.5/share/hadoop/mapreduce/lib/
junit-
4.11.jar:/software/hadoop-2.6.5/share/hadoop/mapreduce/
lib/jackson-mapper-asl-
1.9.13.jar:/software/hadoop-2.6.5/share/hadoop/mapreduce/
lib/guice-servlet-
3.0.jar:/software/hadoop-2.6.5/share/hadoop/mapreduce/lib/
guice-
3.0.jar:/software/hadoop-2.6.5/share/hadoop/mapreduce/lib/
jersey-core-
1.9.jar:/software/hadoop-2.6.5/share/hadoop/mapreduce/lib/
log4j-
1.2.17.jar:/software/hadoop-2.6.5/share/hadoop/mapreduce/
lib/aopalliance-
1.0.jar:/software/hadoop-2.6.5/share/hadoop/mapreduce/
hadoop-mapreduce-examples-
2.6.5.jar:/software/hadoop-2.6.5/share/hadoop/mapreduce/
hadoop-mapreduce-client-jobclient-
2.6.5-tests.jar:/software/hadoop-2.6.5/share/hadoop/mapre-
duce/hadoop-
```

```
mapreduce-client-shuffle-2.6.5.jar:/software/hadoop-2.6.5/share/hadoop/mapreduce/hadoop-mapreduce-client-core-2.6.5.jar:/software/hadoop-2.6.5/share/hadoop/mapreduce/hadoop-mapreduce-client-hs-plugins-2.6.5.jar:/software/hadoop-2.6.5/share/hadoop/mapreduce/hadoop-mapreduce-client-hs-2.6.5.jar:/software/hadoop-2.6.5/share/hadoop/mapreduce/hadoop-mapreduce-client-common-2.6.5.jar:/software/hadoop-2.6.5/share/hadoop/mapreduce/hadoop-mapreduce-client-app-2.6.5.jar:/software/hadoop-2.6.5/share/hadoop/mapreduce/hadoop-mapreduce-client-jobclient-2.6.5.jar:/software/hadoop-2.6.5/contrib/capacity-scheduler/*.jarSTARTUP_MSG: build = https://github.com/apache/hadoop.git -r e8c9fe0b4c252caf2ebf1464220599650f119997; compiled by 'sjlee' on 2016-10-02T23:43Z
STARTUP_MSG: java = 1.8.0_181
************************************************************/
19/04/14 22:24:01 INFO namenode.NameNode: registered UNIX signal handlers for [TERM, HUP, INT]
19/04/14 22:24:01 INFO namenode.NameNode: createNameNode [-bootstrapStandby]
19/04/14 22:24:02 WARN util.NativeCodeLoader: Unable to load native-hadoop library for your platform... using builtin-java classes where applicable
=====================================================
About to bootstrap Standby ID nn2 from:
Nameservice ID: mycluster
Other Namenode ID: nn1
Other NN's HTTP address: http://mynode1:50070
Other NN's IPC address: mynode1/192.168.179.13:8020
Namespace ID: 1476039173
Block pool ID: BP-481747682-192.168.179.13-1555251561577
Cluster ID: CID-90ae45f6-03bd-4c69-8b37-d04b07718e7d
Layout version: -60
```

```
isUpgradeFinalized: true
============================================================
19/04/14 22:24:02 INFO common.Storage: Storage directory /opt/data/hadoop/dfs/name has been successfully formatted.
19/04/14 22:24:03 INFO namenode.TransferFsImage: Opening connection to http://mynode1:50070/imagetransfer?getimage=1&txid=0&storageInfo=-60:1476039173:0:CID-90ae45f6-03bd-4c69-8b37-d04b07718e7d
19/04/14 22:24:03 INFO namenode.TransferFsImage: Image Transfer timeout configured to 60000 milliseconds
19/04/14 22:24:03 INFO namenode.TransferFsImage: Transfer took 0.00s at 0.00 KB/s
19/04/14 22:24:03 INFO namenode.TransferFsImage: Downloaded file fsimage.ckpt_0000000000000000000 size 321 bytes.
19/04/14 22:24:03 INFO util.ExitUtil: Exiting with status 0
19/04/14 22:24:03 INFO namenode.NameNode: SHUTDOWN_MSG:
/************************************************************
SHUTDOWN_MSG: Shutting down NameNode at mynode2/192.168.179.14
************************************************************/
```

### 3.1.3　启动 Hadoop 集群

通过以上配置、格式化及同步 NameNode，现在可以启动 HDFS 集群，启动命令如下：

```
#启动 HDFS 集群
[root@mynode1 ~]# start-dfs.sh
19/04/14 22:25:45 WARN util.NativeCodeLoader: Unable to load native-hadoop library for your platform... using builtin-java classes where applicable
Starting namenodes on [mynode1 mynode2]
mynode1: namenode running as process 1565. Stop it first.
mynode2: starting namenode, logging to /software/hadoop-2.6.5/logs/hadoop-root-namenode-mynode2.out
mynode5: starting datanode, logging to /software/hadoop-2.6.5/logs/hadoop-root-datanode-mynode5.out
mynode3: starting datanode, logging to /software/hadoop-2.6.5/logs/hadoop-root-datanode-mynode3.out
```

```
mynode4: starting datanode, logging to /software/hadoop-2.6.5/logs/hadoop-root-datanode-mynode4.out
Starting journal nodes [mynode3 mynode4 mynode5]
mynode3: journalnode running as process 1450. Stop it first.
mynode5: journalnode running as process 1424. Stop it first.
mynode4: journalnode running as process 1430. Stop it first.
19/04/14 22:25:59 WARN util.NativeCodeLoader: Unable to load native-hadoop library for your platform... using builtin-java classes where applicable
Starting ZK Failover Controllers on NN hosts [mynode1 mynode2]
mynode1: starting zkfc, logging to /software/hadoop-2.6.5/logs/hadoop-root-zkfc-mynode1.out
mynode2: starting zkfc, logging to /software/hadoop-2.6.5/logs/hadoop-root-zkfc-mynode2.out

#启动 Yarn 集群
[root@mynode1 ~]# start-yarn.sh
starting yarn daemons
starting resourcemanager, logging to /software/hadoop-2.6.5/logs/yarn-root-resourcemanager-mynode2.out
mynode5: starting nodemanager, logging to /software/hadoop-2.6.5/logs/yarn-root-nodemanager-mynode5.out
mynode4: starting nodemanager, logging to /software/hadoop-2.6.5/logs/yarn-root-nodemanager-mynode4.out
mynode3: starting nodemanager, logging to /software/hadoop-2.6.5/logs/yarn-root-nodemanager-mynode3.out
```

### 3.1.4 验证 Hadoop 集群节点

分别在 mynode1、mynode2、mynode3、mynode4、mynode5 节点上验证各个进程是否存在，验证 Hadoop 是否搭建成功：

```
#在 mynode1 上执行命令 jps
[root@mynode1 software]# jps
3152 ResourceManager
2789 NameNode
3063 DFSZKFailoverController
3261 Jps
```

```

#在 mynode2 上执行命令 jps
[root@mynode2 ~]# jps
2390 DFSZKFailoverController
2443 Jps
1886 ResourceManager
2302 NameNode

#在 mynode3 上执行命令 jps
[root@mynode3 ~]#jps
2096 NodeManager
1924 DataNode
2006 JournalNode
1388 QuorumPeerMain
2207 Jps

#在 mynode4 上执行命令 jps
[root@mynode4 ~]#jps
1905 DataNode
1987 JournalNode
1380 QuorumPeerMain
2076 NodeManager
2207 Jps

#在 mynode5 上执行命令 jps
[root@mynode5 ~]#jps
1380 QuorumPeerMain
2199 Jps
1898 DataNode
2058 NodeManager
1980 JournalNode
```

对以上节点验证进程是否存在后，Hadoop 集群启动成功。也可以通过验证 Hadoop WEBUI 来验证 Hadoop 集群是否搭建成功：登录 HDFS 节点，在浏览器中输入“http://mynode1:50070”，验证 Nadoop 集群的 WEBUI 界面如图 3－1 所示。

浏览器登录 http://mynode1:8088，验证 Yarn WEBUI 是否正常，如图 3－2 所示。

Hadoop Overview Datanodes Snapshot Startup Progress Utilities

## Overview 'mynode1:8020' (active)

| | |
|---|---|
| Started: | Sun Apr 14 22:34:00 CST 2019 |
| Version: | 2.6.5, re8c9fe0b4c252caf2ebf1464220599650f119997 |
| Compiled: | 2016-10-02T23:43Z by sjlee from branch-2.6.5 |
| Cluster ID: | CID-90ae45f6-03bd-4c69-8b37-d04b07718e7d |
| Block Pool ID: | BP-481747682-192.168.179.13-1555251561577 |

## Summary

Security is off.

Safemode is off.

1 files and directories, 0 blocks = 1 total filesystem object(s).

Heap Memory used 74.5 MB of 153 MB Heap Memory. Max Heap Memory is 889 MB.

Non Heap Memory used 39.2 MB of 40.88 MB Commited Non Heap Memory. Max Non Heap Memory is -1 B.

| | |
|---|---|
| Configured Capacity: | 51.32 GB |
| DFS Used: | 72 KB |

图 3－1 HDFS WEBUI 界面

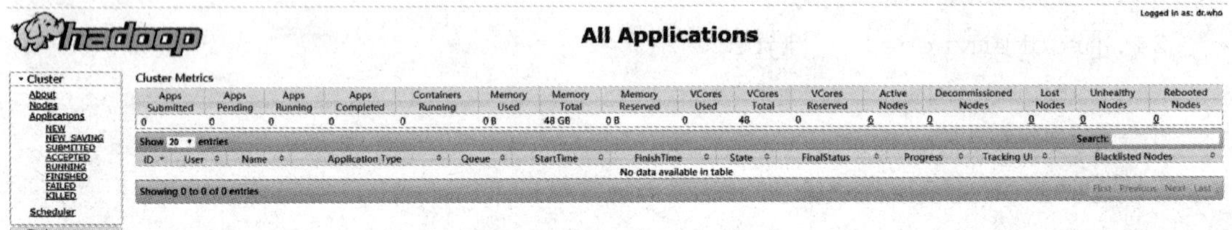


图 3－2 Yarn WEBUI

# 任务 3.2 基于 Yarn 搭建

Spark 任务也可以基于 Yarn 资源调度框架进行资源调度，即 Spark 任务可以提交到 Yarn 上运行。基于 Yarn 提交任务之前，需要做简单的配置，过程如下。

## 3.2.1 节点划分配置

以 mynode4 为 Spark 客户端，在 mynode4 上向 Yarn 资源调度框架提交任务。客户端要提交 Spark 任务，需要在 mynode4 上有 Spark 的安装包，目前在 mynode4 /software 路径下有 Spark 安装包，这样在/software/spark－2.3.1/bin 路径下就可以写入基于 Yarn 提交 Spark 任务的命令。

## 3.2.2 配置 Spark on Yarn

在客户端提交 Spark 任务之前，需要在客户端/software/spark－2.3.1/conf 中配置spark－env.sh 文件，指定 Hadoop 的配置文件路径，这样基于 Yarn 提交任务时，Spark 就可以找到

Yarn 的节点有哪些，从而利用 Yarn 资源调度框架来调度 Spark 程序。配置如下：

```
#编辑修改 spark-env.sh ,配置 Hadoop 文件路径
[root@mynode4 conf]# cd /software/spark-2.3.1/conf/
[root@mynode4 conf]# cp spark-env.sh.template spark-env.sh
[root@mynode4 conf]# vim spark-env.sh

#增加如下内容,保存即可
export HADOOP_CONF_DIR=$HADOOP_HOME/etc/hadoop
```

除了以上配置之外，还需要在 Hadoop 每个 NodeManager 节点的/software/hadoop-2.6.5/etc/hadoop/yarn-site.xml 文件中配置关闭虚拟内存检查。这里 Hadoop 中默认检查虚拟内存，如果实际节点的内存不充足，就会有问题。配置如下：

```
#在 mynode3 节点中配置/software/hadoop-2.6.5/etc/hadoop/yarn-site.xml 文件,增加配置
[root@mynode3 hadoop]# vim /software/hadoop-2.6.5/etc/hadoop/yarn-site.xml
#增加如下配置
<property>
<name>yarn.nodemanager.vmem-check-enabled</name>
<value>false</value>
</property>

#在 mynode4 节点中配置 /software/hadoop-2.6.5/etc/hadoop/yarn-site.xml 文件,增加配置
[root@mynode4 hadoop]# vim /software/hadoop-2.6.5/etc/hadoop/yarn-site.xml
#增加如下配置
<property>
<name>yarn.nodemanager.vmem-check-enabled</name>
<value>false</value>
</property>

#在 mynode5 节点中配置/software/hadoop-2.6.5/etc/hadoop/yarn-site.xml 文件,增加配置
[root@mynode5 hadoop]# vim /software/hadoop-2.6.5/etc/hadoop/yarn-site.xml
#增加如下配置
<property>
<name>yarn.nodemanager.vmem-check-enabled</name>
```

```
<value>false</value>
</property>
```

完成上述配置后，重启 Hadoop 集群，此时，环境准备就绪，在 mynode4 Spark 的客户端上可以基于 Yarn 提交 Spark 任务。

### 3.2.3 验证 Spark on Yarn

基于 Yarn 提交 Spark 任务，仍然选用 Spark 源码中自带的计算 SparkPi 任务例子为例。SparkPi 任务的主类 class 为 org.apache.spark.examples.SparkPi，当前主类所在的 jar 包路径为/software/spark-2.3.1/examples/jars/spark-examples_2.11-2.3.1.jar。Spark 基于 Yarn 提交任务命令、运行任务结果如下：

```
[root@mynode4 conf]# cd /software/spark-2.3.1/bin/
[root@mynode4 bin]# ./spark-submit --master yarn --class org.apache.spark.examples.SparkPi ../examples/jars/spark-examples_2.11-2.3.1.jar
2019-04-14 23:51:48 WARN NativeCodeLoader:62 - Unable to load native-hadoop library for your platform... using builtin-java classes where applicable
2019-04-14 23:51:48 INFO SparkContext:54 - Running Spark version 2.3.1
2019-04-14 23:51:48 INFO SparkContext:54 - Submitted application: Spark Pi
2019-04-14 23:51:48 INFO SecurityManager:54 - Changing view acls to: root
2019-04-14 23:51:48 INFO SecurityManager:54 - Changing modify acls to: root
2019-04-14 23:51:48 INFO SecurityManager:54 - Changing view acls groups to:
2019-04-14 23:51:48 INFO SecurityManager:54 - Changing modify acls groups to:
2019-04-14 23:51:48 INFO SecurityManager:54 - SecurityManager: authentication disabled; uiacls disabled; users with view permissions: Set(root); groups with view permissions: Set(); users with modify permissions: Set(root); groups with modify permissions: Set()
2019-04-14 23:51:49 INFO Utils:54 - Successfully started service 'sparkDriver' on port 35753.
2019-04-14 23:51:49 INFO SparkEnv:54 - Registering MapOutputTracker
```

```
13. 2019-04-14 23:51:49 INFO SparkEnv:54 - Registering BlockMan-
    agerMaster
14. 2019-04-14 23:51:49 INFO BlockManagerMasterEndpoint:54 -
    Using org.apache.spark.storage.DefaultTopologyMapper for get-
    ting topology information
15. 2019-04-14 23:51:49 INFO BlockManagerMasterEndpoint:54 -
    BlockManagerMasterEndpoint up
16. 2019-04-14 23:51:49 INFO DiskBlockManager:54 - Created local
    directory at /tmp/blockmgr-edc5e0b2-095c-4526-8999-
    97608e468d16
17. 2019-04-14 23:51:49 INFO MemoryStore:54 - MemoryStore started
    with capacity 366.3 MB
18. 2019-04-14 23:51:49 INFO SparkEnv:54 - Registering OutputCom-
    mitCoordinator
19. 2019-04-14 23:51:49 INFO log:192 - Logging initialized @1640ms
20. 2019-04-14 23:51:49 INFO Server:346 - jetty-9.3.z-SNAPSHOT
21. 2019-04-14 23:51:49 INFO Server:414 - Started @1719ms
22. 2019-04-14 23:51:49 INFO AbstractConnector:278 - Started
    ServerConnector@3e2fc448{HTTP/1.1,[http/1.1]}{0.0.0.0:4040}
23. 2019-04-14 23:51:49 INFO Utils:54 - Successfully started serv-
    ice 'SparkUI' on port 4040.
24. 2019-04-14 23:51:49 INFO ContextHandler:781 - Started o.s.j.s.
    ServletContextHandler@182b435b{/jobs,null,AVAILABLE,@Spark}
25. 2019-04-14 23:51:49 INFO ContextHandler:781 - Started o.s.j.s.
    ServletContextHandler@12f3afb5{/jobs/json,null,AVAILABLE,
    @Spark}
26. 2019-04-14 23:51:49 INFO ContextHandler:781 - Started o.s.j.s.
    ServletContextHandler@4ced35ed{/jobs/job,null,AVAILABLE,
    @Spark}
27. 2019-04-14 23:51:49 INFO ContextHandler:781 - Started o.s.j.s.
    ServletContextHandler@7bd69e82{/jobs/job/json,null,AVAILA-
    BLE,@Spark}
28. 2019-04-14 23:51:49 INFO ContextHandler:781 - Started o.s.j.s.
    ServletContextHandler@74d7184a{/stages,null,AVAILABLE,
    @Spark}
29. 2019-04-14 23:51:49 INFO ContextHandler:781 - Started o.s.j.s.
    ServletContextHandler@51b01960{/stages/json,null,AVAILABLE,
    @Spark}
```

```
30. 2019-04-14 23:51:49 INFO ContextHandler:781 - Started o.s.j.s.
    ServletContextHandler@6831d8fd{/stages/stage,null,AVAILA-
    BLE,@Spark}
31. 2019-04-14 23:51:49 INFO ContextHandler:781 - Started o.s.j.s.
    ServletContextHandler@3aaf4f07{/stages/stage/json,null,
    AVAILABLE,@Spark}
32. 2019-04-14 23:51:49 INFO ContextHandler:781 - Started o.s.j.s.
    ServletContextHandler@5cbf9e9f{/stages/pool,null,AVAILABLE,
    @Spark}
33. 2019-04-14 23:51:49 INFO ContextHandler:781 - Started o.s.j.s.
    ServletContextHandler@18e8473e{/stages/pool/json,null,
    AVAILABLE,@Spark}
34. 2019-04-14 23:51:49 INFO ContextHandler:781 - Started o.s.j.s.
    ServletContextHandler@5a2f016d{/storage,null,AVAILABLE,
    @Spark}
35. 2019-04-14 23:51:49 INFO ContextHandler:781 - Started o.s.j.s.
    ServletContextHandler@1a38ba58{/storage/json,null,AVAILABLE,
    @Spark}
36. 2019-04-14 23:51:49 INFO ContextHandler:781 - Started o.s.j.s.
    ServletContextHandler@3ad394e6{/storage/rdd,null,AVAILABLE,
    @Spark}
37. 2019-04-14 23:51:49 INFO ContextHandler:781 - Started o.s.j.s.
    ServletContextHandler@6058e535{/storage/rdd/json,null,
    AVAILABLE,@Spark}
38. 2019-04-14 23:51:49 INFO ContextHandler:781 - Started o.s.j.s.
    ServletContextHandler@42deb43a{/environment,null,AVAILABLE,
    @Spark}
39. 2019-04-14 23:51:49 INFO ContextHandler:781 - Started o.s.j.s.Serv-
    letContextHandler@1deb2c43{/environment/json,null,AVAILABLE,
    @Spark}
40. 2019-04-14 23:51:49 INFO ContextHandler:781 - Started o.s.j.s.
    ServletContextHandler@3bb9efbc{/executors,null,AVAILABLE,
    @Spark}
41. 2019-04-14 23:51:49 INFO ContextHandler:781 - Started o.s.j.s.
    ServletContextHandler@1cefc4b3{/executors/json,null,AVAILA-
    BLE,@Spark}
42. 2019-04-14 23:51:49 INFO ContextHandler:781 - Started o.s.j.s.
```

```
    ServletContextHandler@2b27cc70{/executors/threadDump,null,
    AVAILABLE,@Spark}
43. 2019-04-14 23:51:49 INFO ContextHandler:781 - Started o.s.j.s.
    ServletContextHandler@6f6a7463{/executors/threadDump/json,
    null,AVAILABLE,@Spark}
44. 2019-04-14 23:51:49 INFO ContextHandler:781 - Started o.s.j.s.
    ServletContextHandler@1bdaa23d{/static, null, AVAILABLE,
    @Spark}
45. 2019-04-14 23:51:49 INFO ContextHandler:781 - Started o.s.j.s.
    ServletContextHandler@3a45c42a{/,null,AVAILABLE,@Spark}
46. 2019-04-14 23:51:49 INFO ContextHandler:781 - Started o.s.j.s.
    ServletContextHandler@36dce7ed{/api,null,AVAILABLE,@Spark}
47. 2019-04-14 23:51:49 INFO ContextHandler:781 - Started o.s.j.s.
    ServletContextHandler@7692cd34{/jobs/job/kill,null,AVAILA-
    BLE,@Spark}
48. 2019-04-14 23:51:49 INFO ContextHandler:781 - Started o.s.j.s.
    ServletContextHandler@33aa93c{/stages/stage/kill, null,
    AVAILABLE,@Spark}
49. 2019-04-14 23:51:49 INFO SparkUI:54 - Bound SparkUI to 0.0.0.0,
    and started at http://mynode4:4040
50. 2019-04-14 23:51:49 INFO SparkContext:54 - Added JAR file:/
    software/spark-2.3.1/bin/../examples/jars/spark-examples_
    2.11-2.3.1.jar at
    spark://mynode4:35753/jars/spark-examples_2.11-2.3.1.jar
    with timestamp 1555257109679
51. 2019-04-14 23:51:50 INFO ConfiguredRMFailoverProxyProvider:
    100 - Failing over to rm2
52. 2019-04-14 23:51:50 INFO Client:54 - Requesting a new applica-
    tion from cluster with 6 NodeManagers
53. 2019-04-14 23:51:50 INFO Client:54 - Verifying our application
    has not requested more than the maximum memory capability of the
    cluster (8192 MB per container)
54. 2019-04-14 23:51:50 INFO Client:54 - Will allocate AM contain-
    er, with 896 MB memory including 384 MB overhead
55. 2019-04-14 23:51:50 INFO Client:54 - Setting up container
    launch context for our AM
56. 2019-04-14 23:51:50 INFO Client:54 - Setting up the launch
    environment for our AM container
```

```
2019-04-14 23:51:50 INFO Client:54 - Preparing resources for our AM container
2019-04-14 23:51:51 WARN Client:66 - Neither spark.yarn.jars nor spark.yarn.archive is set, falling back to uploading libraries under SPARK_HOME.
2019-04-14 23:51:52 INFO Client:54 - Uploading resource file:/tmp/spark-30e102ab-f14d-41e0-b491-30f94249e1a7/__spark_libs__7445946215708104422.zip -> hdfs://mycluster/user/root/.sparkStaging/application_1555252000899_0002/__spark_libs__7445946215708104422.zip
2019-04-14 23:51:55 INFO Client:54 - Uploading resource file:/tmp/spark-30e102ab-f14d-41e0-b491-30f94249e1a7/__spark_conf__3951449412469096266.zip -> hdfs://mycluster/user/root/.sparkStaging/application_1555252000899_0002/__spark_conf__.zip
2019-04-14 23:51:55 INFO SecurityManager:54 - Changing view acls to: root
2019-04-14 23:51:55 INFO SecurityManager:54 - Changing modify acls to: root
2019-04-14 23:51:55 INFO SecurityManager:54 - Changing view acls groups to:
2019-04-14 23:51:55 INFO SecurityManager:54 - Changing modify acls groups to:
2019-04-14 23:51:55 INFO SecurityManager:54 - SecurityManager: authentication disabled; ui acls disabled; users with view permissions: Set(root); groups with view permissions: Set(); users with modify permissions: Set(root); groups with modify permissions: Set()
2019-04-14 23:51:55 INFO Client:54 - Submitting application application_1555252000899_0002 to ResourceManager
2019-04-14 23:51:55 INFO YarnClientImpl:251 - Submitted application application_1555252000899_0002
2019-04-14 23:51:55 INFO SchedulerExtensionServices:54 - Starting Yarn extension services with app application_1555252000899_0002 and attemptId None
2019-04-14 23:51:56 INFO Client:54 - Application report for application_1555252000899_0002 (state: ACCEPTED)
2019-04-14 23:51:56 INFO Client:54 -
```

```
client token: N/A
diagnostics: N/A
ApplicationMaster host: N/A
ApplicationMaster RPC port: -1
queue: default
start time: 1555257116017
final status: UNDEFINED
tracking URL: http://mynode2:8088/proxy/application_1555252000899_0002/
user: root
2019-04-14 23:51:58 INFO Client:54 - Application report for application_1555252000899_0002 (state: ACCEPTED)
2019-04-14 23:51:59 INFO Client:54 - Application report for application_1555252000899_0002 (state: ACCEPTED)
2019-04-14 23:52:00 INFO Client:54 - Application report for application_1555252000899_0002 (state: ACCEPTED)
2019-04-14 23:52:01 INFO Client:54 - Application report for application_1555252000899_0002 (state: ACCEPTED)
2019-04-14 23:52:02 INFO Client:54 - Application report for application_1555252000899_0002 (state: ACCEPTED)
2019-04-14 23:52:03 INFO Client:54 - Application report for application_1555252000899_0002 (state: ACCEPTED)
2019-04-14 23:52:03 INFO YarnClientSchedulerBackend:54 - Add WebUI Filter. org.apache.hadoop.yarn.server.webproxy.amfilter.AmIpFilter, Map(PROXY_HOSTS -> mynode1,mynode2, PROXY_URI_BASES -> http://mynode1:8088/proxy/application_1555252000899_0002,http://mynode2:8088/proxy/application_1555252000899_0002), /proxy/application_1555252000899_0002
2019-04-14 23:52:03 INFO JettyUtils:54 - Adding filter: org.apache.hadoop.yarn.server.webproxy.amfilter.AmIpFilter
2019-04-14 23:52:03 INFO YarnSchedulerBackend$YarnSchedulerEndpoint:54 - ApplicationMaster registered as NettyRpcEndpointRef(spark-client://YarnAM)
2019-04-14 23:52:04 INFO Client:54 - Application report for application_1555252000899_0002 (state: RUNNING)
2019-04-14 23:52:04 INFO Client:54 -
client token: N/A
```

```
diagnostics: N/A
ApplicationMaster host: 192.168.179.17
ApplicationMaster RPC port: 0
queue: default
start time: 1555257116017
final status: UNDEFINED
tracking URL: http://mynode2:8088/proxy/application_1555252000899_0002/
user: root
2019-04-14 23:52:04 INFO YarnClientSchedulerBackend:54 - Application application_1555252000899_0002 has started running.
2019-04-14 23:52:04 INFO Utils:54 - Successfully started service 'org.apache.spark.network.netty.NettyBlockTransferService' on port 54515.
2019-04-14 23:52:04 INFO NettyBlockTransferService:54 - Server created on mynode4:54515
2019-04-14 23:52:04 INFO BlockManager:54 - Using org.apache.spark.storage.RandomBlockReplicationPolicy for block replication policy
2019-04-14 23:52:04 INFO BlockManagerMaster:54 - Registering BlockManagerBlockManagerId(driver, mynode4, 54515, None)
2019-04-14 23:52:04 INFO BlockManagerMasterEndpoint:54 - Registering block manager mynode4:54515 with 366.3 MB RAM, BlockManagerId(driver, mynode4, 54515, None)
2019-04-14 23:52:04 INFO BlockManagerMaster:54 - Registered BlockManagerBlockManagerId(driver, mynode4, 54515, None)
2019-04-14 23:52:04 INFO BlockManager:54 - Initialized BlockManager: BlockManagerId(driver, mynode4, 54515, None)
2019-04-14 23:52:04 INFO ContextHandler:781 - Started o.s.j.s.ServletContextHandler@68af87ad{/metrics/json,null,AVAILABLE,@Spark}
2019-04-14 23:52:08 INFO YarnSchedulerBackend$YarnDriverEndpoint:54 - Registered executor NettyRpcEndpointRef(spark-client://Executor) (192.168.179.17:38190) with ID 1
2019-04-14 23:52:08 INFO BlockManagerMasterEndpoint:54 - Registering block manager mynode5:48994 with 366.3 MB RAM, BlockManagerId(1, mynode5, 48994, None)
```

```
111. 2019-04-14 23:52:10 INFO YarnSchedulerBackend$YarnDriverEndpoint:54 - Registered executor NettyRpcEndpointRef(spark-client://Executor) (192.168.179.16:56652) with ID 2
112. 2019-04-14 23:52:10 INFO YarnClientSchedulerBackend:54 - SchedulerBackend is ready for scheduling beginning after reached minRegisteredResourcesRatio: 0.8
113. 2019-04-14 23:52:10 INFO BlockManagerMasterEndpoint:54 - Registering block manager mynode4:44521 with 366.3 MB RAM, BlockManagerId(2, mynode4, 44521, None)
114. 2019-04-14 23:52:10 INFO SparkContext:54 - Starting job: reduce at SparkPi.scala:38
115. 2019-04-14 23:52:10 INFO DAGScheduler:54 - Got job 0 (reduce at SparkPi.scala:38) with 2 output partitions
116. 2019-04-14 23:52:10 INFO DAGScheduler:54 - Final stage: ResultStage 0 (reduce at SparkPi.scala:38)
117. 2019-04-14 23:52:10 INFO DAGScheduler:54 - Parents of final stage: List()
118. 2019-04-14 23:52:10 INFO DAGScheduler:54 - Missing parents: List()
119. 2019-04-14 23:52:10 INFO DAGScheduler:54 - Submitting ResultStage 0 (MapPartitionsRDD[1] at map at SparkPi.scala:34), which has no missing parents
120. 2019-04-14 23:52:10 INFO MemoryStore:54 - Block broadcast_0 stored as values in memory (estimated size 1832.0 B, free 366.3 MB)
121. 2019-04-14 23:52:10 INFO MemoryStore:54 - Block broadcast_0_piece0 stored as bytes in memory (estimated size 1181.0 B, free 366.3 MB)
122. 2019-04-14 23:52:10 INFO BlockManagerInfo:54 - Added broadcast_0_piece0 in memory on mynode4:54515 (size: 1181.0 B, free: 366.3 MB)
123. 2019-04-14 23:52:10 INFO SparkContext:54 - Created broadcast 0 from broadcast at DAGScheduler.scala:1039
124. 2019-04-14 23:52:10 INFO DAGScheduler:54 - Submitting 2 missing tasks from ResultStage 0 (MapPartitionsRDD[1] at map at SparkPi.scala:34) (first 15 tasks are for partitions Vector(0, 1))
125. 2019-04-14 23:52:10 INFO YarnScheduler:54 - Adding task set 0.0 with 2 tasks
```

```
2019-04-14 23:52:10 INFO TaskSetManager:54 - Starting task 0.0 in stage 0.0 (TID 0, mynode4, executor 2, partition 0, PROCESS_LOCAL, 7864 bytes)
2019-04-14 23:52:10 INFO TaskSetManager:54 - Starting task 1.0 in stage 0.0 (TID 1, mynode5, executor 1, partition 1, PROCESS_LOCAL, 7864 bytes)
2019-04-14 23:52:11 INFO BlockManagerInfo:54 - Added broadcast_0_piece0 in memory on mynode5:48994 (size: 1181.0 B, free: 366.3 MB)
2019-04-14 23:52:11 INFO TaskSetManager:54 - Finished task 1.0 in stage 0.0 (TID 1) in 658 ms on mynode5 (executor 1) (1/2)
2019-04-14 23:52:11 INFO BlockManagerInfo:54 - Added broadcast_0_piece0 in memory on mynode4:44521 (size: 1181.0 B, free: 366.3 MB)
2019-04-14 23:52:11 INFO TaskSetManager:54 - Finished task 0.0 in stage 0.0 (TID 0) in 951 ms on mynode4 (executor 2) (2/2)
2019-04-14 23:52:11 INFO DAGScheduler:54 - ResultStage 0 (reduce at SparkPi.scala:38) finished in 1.359 s
2019-04-14 23:52:11 INFO YarnScheduler:54 - Removed TaskSet 0.0, whose tasks have all completed, from pool
2019-04-14 23:52:11 INFO DAGScheduler:54 - Job 0 finished: reduce at SparkPi.scala:38, took 1.440064 s
Pi is roughly 3.1385556927784637 #Spark Pi 结果
2019-04-14 23:52:11 INFO AbstractConnector:318 - Stopped Spark@3e2fc448{HTTP/1.1,[http/1.1]}{0.0.0.0:4040}
2019-04-14 23:52:11 INFO SparkUI:54 - Stopped Spark web UI at http://mynode4:4040
2019-04-14 23:52:11 INFO YarnClientSchedulerBackend:54 - Interrupting monitor thread
2019-04-14 23:52:11 INFO YarnClientSchedulerBackend:54 - Shutting down all executors
2019-04-14 23:52:11 INFO YarnSchedulerBackend$YarnDriverEndpoint:54 - Asking each executor to shut down
2019-04-14 23:52:11 INFO SchedulerExtensionServices:54 - Stopping SchedulerExtensionServices
(serviceOption=None,
services=List(),
started=false)
```

```
145. 2019 - 04 - 14 23:52:11 INFO YarnClientSchedulerBackend:54 -
     Stopped
146. 2019 - 04 - 14 23:52:12 INFO MapOutputTrackerMasterEndpoint:54 -
     MapOutputTrackerMasterEndpoint stopped!
147. 2019 - 04 - 14 23:52:12 INFO MemoryStore:54 - MemoryStore cleared
148. 2019 - 04 - 14 23:52:12 INFO BlockManager:54 - BlockManager stopped
149. 2019 - 04 - 14 23:52:12 INFO BlockManagerMaster:54 - BlockMan-
     agerMaster stopped
150. 2019 - 04 - 14 23:52:12 INFO OutputCommitCoordinator$OutputCom-
     mitCoordinatorEndpoint:54 - OutputCommitCoordinator stopped!
151. 2019 - 04 - 14 23:52:12 INFO SparkContext:54 - Successfully
     stopped SparkContext
152. 2019 - 04 - 14 23:52:12 INFO ShutdownHookManager:54 - Shutdown
     hook called
153. 2019 - 04 - 14 23:52:12 INFO ShutdownHookManager:54 - Deleting di-
     rectory /tmp/spark - 30e102ab - f14d - 41e0 - b491 - 30f94249e1a7
154. 2019 - 04 - 14 23:52:12 INFO ShutdownHookManager:54 - Deleting di-
     rectory /tmp/spark - a1243066 - 0b56 - 40ab - a2d0 - 5df8598ee49c
```

通过以上运行结果可知，Spark 基于 Yarn 运行任务成功，可以基于 Yarn 资源调度框架运行 ×× 系统。

# 项目四
# 日志服务器搭建

【项目描述】

在××系统中，当将任务提交到集群中运行时，需要在集群中查看任务运行的日志，这样就需要配置 Spark 的日志管理，这就是本项目需要解决的问题。

【项目分析】

无论是在 Spark 自带的 Standalone 集群中运行 Spark 任务，还是在Hadoop 生态圈中的 Yarn 上运行 Spark 任务，除了需要知道各自的配置及任务提交命令，还需要知道如何查看当前任务的日志及历史应用程序的日志，这样有助于查找任务运行的状态，不断优化任务。

## 任务 4.1　日志服务器配置

以基于 Hadoop 生态圈中的 Yarn 为例，介绍如何配置才可以在集群中查看 Spark 任务日志、如何配置才可以查看运行过的应用程序的日志，这样以后××系统在集群中运行时，可以查看日志，帮助优化程序。

Spark 的任务基于 Yarn 提交运行时，每个 Spark 应用程序启动之后都会启动一个 Driver JVM 进程，这个进程的作用是向 Yarn 资源调度框架中的 NodeManager 发送 task。Driver 可以在提交任务的客户端启动。基于 Yarn 提交任务的命令如下：

```
8../spark-submit --master yarn --class 主类 主类所在包的位置
```

提交以上命令，Driver 将在客户端启动，这样如果服务器不能将客户端 Driver 运行的日志收集起来，就不方便管理。为了方便统一管理 Spark 任务运行的所有日志，可以将 Driver 运行在 Yarn 的一台 NodeManager 节点上。使用以下命令基于 Yarn 资源调度框架提交 Spark 任务。

```
./spark-submit --master yarn --deploy-mode cluster --class 主类 主类所在包的位置
```

通过以上配置，可以搭建 Spark 的日志管理系统，将 Spark 的所有日志进行保存和查看。

下面以基于 Yarn 提交 SparkPi 的计算案例为例来说明如何配置 Spark 基于 Yarn 的日志管理与查看。

首先，启动 Hadoop 集群，在 mynode1 执行命令 start-dfs.sh、start-yarn.sh 启动集群：

```
[root@mynode1 ~]# start-dfs.sh
19/04/1723:56:18 WARN util.NativeCodeLoader: Unable to load native-hadoop library for your platform...usingbuiltin-java classes where applicable
Startingnamenodes on [mynode1 mynode2]
mynode1: starting namenode, logging to /software/hadoop-2.6.5/logs/hadoop-root-namenode-mynode1.out
mynode2: starting namenode, logging to /software/hadoop-2.6.5/logs/hadoop-root-namenode-mynode2.out
mynode4: starting datanode, logging to /software/hadoop-2.6.5/logs/hadoop-root-datanode-mynode4.out
mynode3: starting datanode, logging to /software/hadoop-2.6.5/logs/hadoop-root-datanode-mynode3.out
mynode5: starting datanode, logging to /software/hadoop-2.6.5/logs/hadoop-root-datanode-mynode5.out
Starting journal nodes [mynode3 mynode4 mynode5]
mynode3: starting journalnode, logging to /software/hadoop-2.6.5/logs/hadoop-root-journalnode-mynode3.out
mynode5: starting journalnode, logging to /software/hadoop-2.6.5/logs/hadoop-root-journalnode-mynode5.out
mynode4: starting journalnode, logging to /software/hadoop-2.6.5/logs/hadoop-root-journalnode-mynode4.out
19/04/1723:56:35 WARN util.NativeCodeLoader: Unable to load native-hadoop library for your platform...usingbuiltin-java classes where applicable
Starting ZK FailoverControllers on NN hosts [mynode1 mynode2]
mynode1: starting zkfc, logging to /software/hadoop-2.6.5/logs/hadoop-root-zkfc-mynode1.out
```

```
mynode2: starting zkfc, logging to /software/hadoop-2.6.5/logs/hadoop-root-zkfc-mynode2.out
[root@mynode1 ~]# start-yarn.sh
starting yarn daemons
starting resourcemanager, logging to /software/hadoop-2.6.5/logs/yarn-root-resourcemanager-mynode1.out
mynode3: starting nodemanager, logging to /software/hadoop-2.6.5/logs/yarn-root-nodemanager-mynode3.out
mynode4: starting nodemanager, logging to /software/hadoop-2.6.5/logs/yarn-root-nodemanager-mynode4.out
mynode5: starting nodemanager, logging to /software/hadoop-2.6.5/logs/yarn-root-nodemanager-mynode5.out
```

然后，在 Spark 的客户端 mynode4 的/software/spark-2.3.1/bin 路径下提交 SparkPi 程序，提交命令如下：

```
[root@mynode4 bin]# cd /software/spark-2.3.1/bin/
[root@mynode4 bin]#./spark-submit --master yarn --deploy-mode cluster --classorg.apache.spark.examples.SparkPi ../examples/jars/spark-examples_2.11-2.3.1.jar
```

在浏览器中打开 Yarn 的任务管理页面 https://mynode1:8088，查看提交任务的运行状态，如图 4-1 所示。

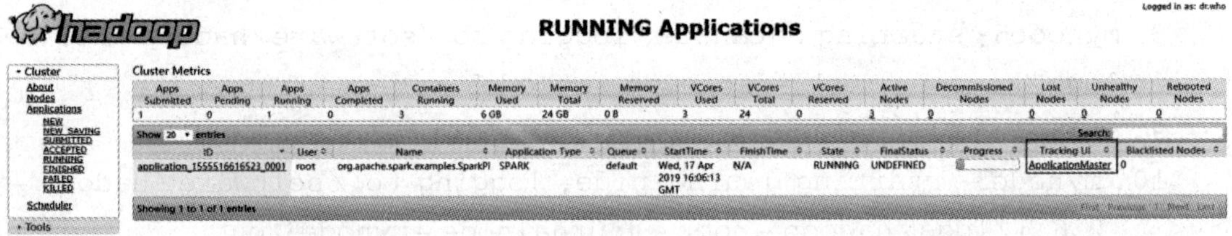


图 4-1　Yarn Application 管理界面

单击图 4-1 中的“Tracking UI”，查看当前运行时任务的状态，如图 4-2 所示。

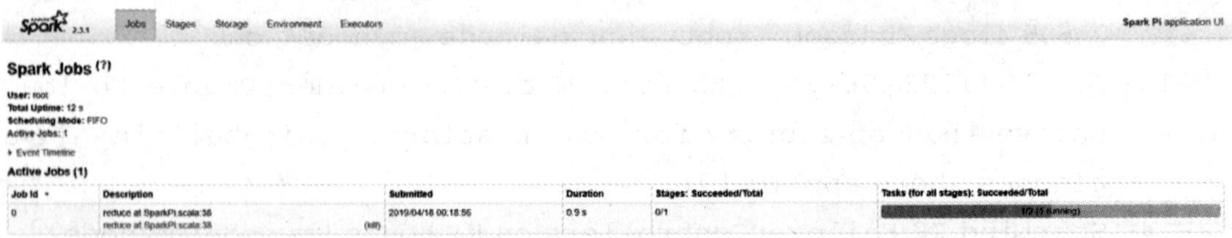


图 4-2　Spark 任务运行时日志

在客户端可以看到基于 Yarn 提交任务后显示的状态信息如下：

```
1. [root@mynode4 bin]#./spark - submit  - - master yarn  - - deploy -
   mode cluster  - - classorg.apache.spark.examples.SparkPi../exam-
   ples/jars/spark - examples_2.11 - 2.3.1.jar
2. 2019 - 04 - 1800:18:30 WARN NativeCodeLoader:62 - Unable to load n-
   ative - hadoop library for your platform...usingbuiltin - java
   classes where applicable
3. 2019 - 04 - 1800:18:31 INFO Client:54 - Requesting a new applica-
   tion from cluster with3NodeManagers
4. 2019 - 04 - 1800:18:31 INFO Client:54 - Verifyingour application
   has not requested more than the maximum memory capability of the
   cluster (8192 MB per container)
5. 2019 - 04 - 1800:18:31 INFO Client:54 - Will allocate AM container,
   with1408 MB memory including 384 MB overhead
6. 2019 - 04 - 1800:18:31 INFO Client:54 - Setting up container launch
   context forour AM
7. 2019 - 04 - 1800:18:31 INFO Client:54 - Setting up the launch envi-
   ronment forour AM container
8. 2019 - 04 - 1800:18:31 INFO Client:54 - Preparing resources forour
   AM container
9. 2019 - 04 - 1800:18:32 WARN Client:66 - Neitherspark.yarn.jars nor
   spark.yarn.archiveisset, falling back to uploading libraries un-
   der SPARK_HOME.
10. 2019 - 04 - 1800:18:33 INFO Client:54 - Uploading resource file:/
    tmp/spark - 7dc9435e - 83c5 - 48c5 - bc12 - 0520965ab0a3/__spark_
    libs__2683705141558867003.zip -> hdfs://mycluster/user/root/.
    sparkStaging/application_1555517861472_0001/__spark_libs__
    2683705141558867003.zip
11. 2019 - 04 - 1800:18:37 INFO Client:54 - Uploading resource file:/
    software/spark - 2.3.1/examples/jars/spark - examples_2.11 - 2.3.
    1.jar -> hdfs://mycluster/user/root/.sparkStaging/application_
    1555517861472_0001/spark - examples_2.11 - 2.3.1.jar
12. 2019 - 04 - 1800:18:37 INFO Client:54 - Uploading resource file:/
    tmp/spark - 7dc9435e - 83c5 - 48c5 - bc12 - 0520965ab0a3/__spark_
    conf__4658795681728934509.zip -> hdfs://mycluster/user/
    root/.sparkStaging/application_1555517861472_0001/__spark_
    conf__.zip
13. 2019 - 04 - 1800:18:37 INFO SecurityManager:54 - Changing view acls
```

```
to: root
2019 -04 -1800:18:37 INFO SecurityManager:54 - Changing modify acls to: root
2019 -04 -1800:18:37 INFO SecurityManager:54 - Changing view acls groups to:
2019 -04 -1800:18:37 INFO SecurityManager:54 - Changing modify acls groups to:
2019 -04 -1800:18:37 INFO SecurityManager:54 - SecurityManager: authentication disabled; uiacls disabled; users with view permissions:Set(root); groups with view permissions:Set(); users with modify permissions:Set(root); groups with modify permissions:Set()
2019 -04 -1800:18:37 INFO Client:54 - Submitting application application_1555517861472_0001 to ResourceManager
2019 -04 -1800:18:37 INFO YarnClientImpl:251 - Submitted application application_1555517861472_0001
2019 -04 -1800:18:38 INFO Client:54 - Application report for application_1555517861472_0001 (state: ACCEPTED)
2019 -04 -1800:18:38 INFO Client:54 -
client token: N/A
diagnostics: N/A
ApplicationMaster host: N/A
ApplicationMaster RPC port: -1
queue:default
start time:1555517917450
final status: UNDEFINED
tracking URL: http://mynode1:8088/proxy/application_1555517861472_0001/
user: root
2019 -04 -1800:18:39 INFO Client:54 - Application report for application_1555517861472_0001 (state: ACCEPTED)
2019 -04 -1800:18:40 INFO Client:54 - Application report for application_1555517861472_0001 (state: ACCEPTED)
2019 -04 -1800:18:41 INFO Client:54 - Application report for application_1555517861472_0001 (state: ACCEPTED)
2019 -04 -1800:18:42 INFO Client:54 - Application report for application_1555517861472_0001 (state: ACCEPTED)
```

```
2019-04-1800:18:43 INFO Client:54 - Application report for application_1555517861472_0001 (state: ACCEPTED)
2019-04-1800:18:44 INFO Client:54 - Application report for application_1555517861472_0001 (state: ACCEPTED)
2019-04-1800:18:45 INFO Client:54 - Application report for application_1555517861472_0001 (state: ACCEPTED)
2019-04-1800:18:46 INFO Client:54 - Application report for application_1555517861472_0001 (state: ACCEPTED)
2019-04-1800:18:47 INFO Client:54 - Application report for application_1555517861472_0001 (state: RUNNING)
2019-04-1800:18:47 INFO Client:54 -
client token: N/A
diagnostics: N/A
ApplicationMaster host:192.168.179.17
ApplicationMaster RPC port:0
queue:default
start time:1555517917450
final status: UNDEFINED
tracking URL: http://mynode1:8088/proxy/application_1555517861472_0001/
user: root
2019-04-1800:18:48 INFO Client:54 - Application report for application_1555517861472_0001 (state: RUNNING)
2019-04-1800:18:49 INFO Client:54 - Application report for application_1555517861472_0001 (state: RUNNING)
2019-04-1800:18:50 INFO Client:54 - Application report for application_1555517861472_0001 (state: RUNNING)
2019-04-1800:18:51 INFO Client:54 - Application report for application_1555517861472_0001 (state: RUNNING)
2019-04-1800:18:52 INFO Client:54 - Application report for application_1555517861472_0001 (state: RUNNING)
2019-04-1800:18:53 INFO Client:54 - Application report for application_1555517861472_0001 (state: RUNNING)
2019-04-1800:18:54 INFO Client:54 - Application report for application_1555517861472_0001 (state: RUNNING)
2019-04-1800:18:55 INFO Client:54 - Application report for application_1555517861472_0001 (state: RUNNING)
```

```
2019 - 04 - 1800:18:56 INFO Client:54 - Application report for
    application_1555517861472_0001 (state: RUNNING)
2019 - 04 - 1800:18:57 INFO Client:54 - Application report for
    application_1555517861472_0001 (state: FINISHED)
2019 -04 -1800:18:57 INFO Client:54 -
client token: N/A
diagnostics: N/A
ApplicationMaster host:192.168.179.17
ApplicationMaster RPC port:0
queue:default
start time:1555517917450
final status: SUCCEEDED
tracking URL: http://mynode1:8088/proxy/application_155551786
    1472_0001/
user: root
2019 - 04 - 1800:18:57 INFO ShutdownHookManager:54 - Shutdown
    hook called
2019 - 04 - 1800:18:57 INFO ShutdownHookManager:54 - Deleting
    directory /tmp/spark - 25d0820a - 4f24 - 4d1c - a7ee - 3bf4d4
    d1e799
2019 - 04 - 1800:18:57 INFO ShutdownHookManager:54 - Deleting
    directory /tmp/spark - 7dc9435e - 83c5 - 48c5 - bc12 - 0520965
    ab0a3
```

完成上述任务后，如果想再次查看任务运行时的日志，在 Yarn 的管理页面中单击“Tracking UI”对应的“History”，弹出如图 4 - 3 所示页面，可以查看当时运行任务的 task 数量、运行任务时占用的内存等。

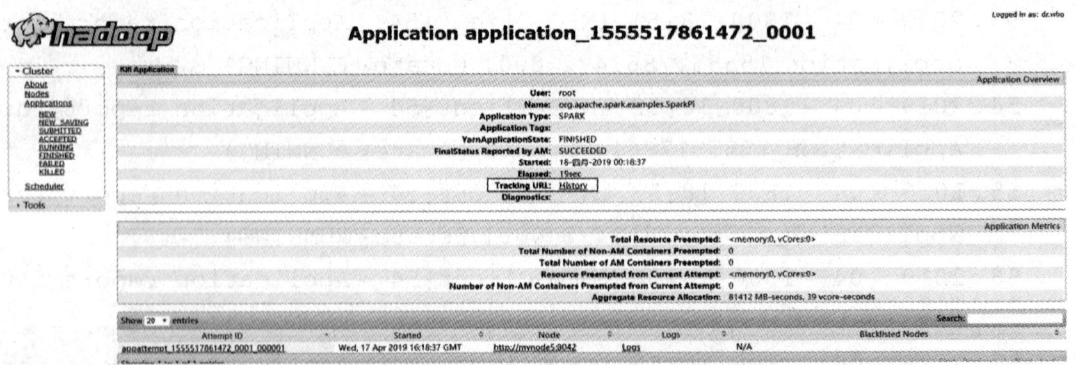


图 4 - 3　查看历史日志

单击“Tracking URL”，发现依然查询不到对应的Spark运行时的状态信息。可以通过配置Yarn任务管理页面的“Tracking URL”的链接指向及Spark的历史日志服务器，来查看已完成的Spark任务的日志问题。

解决上述问题，需要两个步骤：

①配置Yarn Application中已运行完成任务的“Tracking URL”链接信息。

②配置Spark的历史日志服务器。

下面在HDFS和Spark中修改配置，实现以上两个步骤。

### 4.1.1　HDFS配置

在HDFS集群所有节点的yarn - site. xml中追加配置如下信息：

```
<!--开启聚合日志 -->
<property>
<name>yarn.log-aggregation-enable</name>
<value>true</value>
</property>
<!--配置Hadoop日志服务器的地址,查看Spark Executor日志时使用 -->
<property>
<name>yarn.log.server.url</name>
<value>http://mynode1:19888/jobhistory/logs/</value>
</property>
<!--配置日志过期时间,单位为秒 -->
<property>
<name>yarn.log-aggregation.retain-seconds</name>
<value>86400</value>
</property>
```

基于Yarn提交Spark应用程序后，在Spark的运行过程中，NodeManager的日志信息会被Hadoop的jobhistory记录，这样可以找到对应的jobhistory来查看Spark任务的日志。

对应的配置详细解释如下：

yarn. log - aggregation - enable：开启日志的聚合功能，将NodeManager上的运行日志信息进行聚合保存。

yarn. log. server. url：访问NodeManager节点上的日志路径。

yarn. log - aggregation. retain - seconds：配置日志的过期时间。

上述配置完成之后，日志服务器中关于Hadoop的配置都已经完成。需要注意的是，将Hadoop所有节点的yarn - site. xml都做相同的配置。

### 4.1.2　Spark配置

配置历史日志服务器时，需要在提交Spark任务的客户端上做一些简单的设置。这里提交Spark任务的客户端是mynode4，在mynode4的/software/spark-2.3.1/conf路径下配置spark-defaults.conf文件，操作如下：

```
[root@mynode4 conf]# cp spark-defaults.conf.template spark-defaults.conf
[root@mynode4 conf]# vim spark-defaults.conf
```

spark-defaults.conf的配置详细内容如下：

```
# Licensed to the Apache Software Foundation (ASF) under one or more
# contributor license agreements. See the NOTICE file distributed with
# this work for additional information regarding copyright ownership.
# The ASF licenses this file to You under the Apache License, Version 2.0
# (the "License"); you may not use this file except in compliance with
# the License. You may obtain a copy of the License at
#
# http://www.apache.org/licenses/LICENSE-2.0
#
# Unless required by applicable law or agreed to in writing, software
# distributed under the License is distributed on an "AS IS" BASIS,
# WITHOUT WARRANTIES OR CONDITIONS OF ANY KIND, either express or implied.
# See the License for the specific language governing permissions and
# limitations under the License.
#


# Default system properties included when running spark-submit.
# This is useful for setting default environmental settings.


spark.eventLog.enabledtrue#开启 Spark 的日志管理
spark.eventLog.compresstrue#设置 Spark 的日志为压缩格式,这样节省空间
spark.eventLog.dir hdfs:/mycluster/spark/log #设置 Spark 日志的存储位置
```

```
spark.history.fs.logDirectory hdfs://mycluster/spark/log
    #设置 Spark 历史日志服务器恢复日志的路径
spark.yarn.historyServer.address mynode4:18080
    #当在 Yarn 上查看历史日志时,链接到的路径为 mynode4:18080
```

以上配置的详细解释如下：

spark. eventLog. enabled：开启 Spark 的日志管理。

spark. eventLog. compress：将 Spark 的日志进行压缩保存。

spark. eventLog. dir：设置 Spark 应用程序运行后，将 Spark 的日志存在哪个路径下。如将日志信息保存到 HDFS 中，则这个路径需要提前手动在 HDFS 中创建。

spark. history. fs. logDirectory：设置当启动 Spark 的日志服务器时，应该从哪个路径下恢复运行完成的 Spark 程序的完整日志。这个路径信息应该与 spark. eventLog. dir 路径保持一致。

spark. yarn. historyServer. address：设置的是，当在 Yarn Application 管理页面中单击某个已运行完成的 Spark 应用程序的“Tracking UI”时，对应跳转的日志地址。

此时，日志服务器中关于 Spark 的配置已经完成。

### 4.1.3　启动日志服务器

上述配置完成之后，按照以下步骤启动日志服务器。

首先需要在 mynode1 上重启 Hadoop 集群，这样保证在 Hadoop 上的配置信息生效。命令如下：

```
[root@mynode1 ~]# stop-dfs.sh
19/04/18 07:33:08 WARN util.NativeCodeLoader: Unable to load
  native-hadoop library for your platform... using builtin-java
  classes where applicable
Stopping namenodes on [mynode1 mynode2]
mynode2: no namenode to stop
mynode1: stopping namenode
mynode5: stopping datanode
mynode3: stopping datanode
mynode4: stopping datanode
Stopping journal nodes [mynode3 mynode4 mynode5]
mynode4: stopping journalnode
mynode5: stopping journalnode
mynode3: stopping journalnode
19/04/18 07:33:27 WARN util.NativeCodeLoader: Unable to load
  native-hadoop library for your platform... using builtin-java
  classes where applicable
```

```
Stopping ZK FailoverControllers on NN hosts [mynode1 mynode2]
mynode2: stopping zkfc
mynode1: stopping zkfc
[root@mynode1 ~]# stop-yarn.sh
stopping yarn daemons
stopping resourcemanager
mynode3: stopping nodemanager
mynode5: stopping nodemanager
mynode4: stopping nodemanager
noproxyserver to stop
```

当在 Yarn Application 管理页面中查看 Spark 任务时，关于 Spark 在每个 NodeManager 上运行的详细信息，可以链接到 Yarn 的历史日志服务器中详细查看。所以这里需要启动 Yarn 的历史日志服务器，将 NodeManager 运行的日志信息进行保存，启动命令如下：

```
mr-jobhistory-daemon.sh start historyserver
```

最后，还需要在提交 Spark 的客户端节点上启动 Spark 的历史日志服务器，这样保证在提交 Spark 任务的过程中，Spark 的日志可以保存到设置的 HDFS 路径中。启动 Spark 历史日志服务器的命令及启动之后查看进程的操作如下：

```
#在 mynode4 客户端节点中启动 Spark 的历史日志服务器
[root@mynode4 bin]# cd /software/spark-2.3.1/sbin/
[root@mynode4 sbin]#./start-history-server.sh
starting org.apache.spark.deploy.history.HistoryServer, logging to /software/spark-2.3.1/logs/spark-root-org.apache.spark.deploy.history.HistoryServer-1-mynode4.out

#检查历史日志服务器是否启动
[root@mynode4 sbin]#jps
1553NodeManager
1362DataNode
1298QuorumPeerMain
1763Jps
1452JournalNode
1711HistoryServer#此进程就是 Spark 历史日志服务器的进程
```

通过以上步骤，Spark 的历史日志服务器已经启动，可以保存提交到 Yarn 资源调度框架上的 Spark 任务的日志信息。

### 4.1.4 查看日志

可以在提交 Spark 任务的客户端 mynode4 上提交 SparkPi 任务，测试是否可以将 Spark 任务的日志保存成功、是否可以正常查看已经运行完成的任务的日志信息。在 Spark 提交任务的客户端 mynode4 上执行提交任务的命令：

```
1. [root@mynode4 sbin]# cd /software/spark-2.3.1/bin/
2. [root@mynode4 bin]#./spark-submit --master yarn --deploy-mode cluster --classorg.apache.spark.examples.SparkPi../examples/jars/spark-examples_2.11-2.3.1.jar
```

在 mynode4 客户端节点查看任务运行信息：

```
1. [root@mynode4 bin]#./spark-submit --master yarn --deploy-mode cluster --classorg.apache.spark.examples.SparkPi../examples/jars/spark-examples_2.11-2.3.1.jar
2. 2019-04-1821:49:26 WARN NativeCodeLoader:62 - Unable to load native-hadoop library for your platform... usingbuiltin-java classes where applicable
3. 2019-04-1821:49:33 INFO Client:54 - Requesting a new application from cluster with3NodeManagers
4. 2019-04-1821:49:34 INFO Client:54 - Verifyingour application has not requested more than the maximum memory capability of the cluster (8192 MB per container)
5. 2019-04-1821:49:34 INFO Client:54 - Will allocate AM container, with1408 MB memory including 384 MB overhead
6. 2019-04-1821:49:34 INFO Client:54 - Setting up container launch context forour AM
7. 2019-04-1821:49:34 INFO Client:54 - Setting up the launch environment forour AM container
8. 2019-04-1821:49:34 INFO Client:54 - Preparing resources forour AM container
9. 2019-04-1821:49:35 WARN Client:66 - Neitherspark.yarn.jars nor spark.yarn.archiveisset, falling back to uploading libraries under SPARK_HOME.
10. 2019-04-1821:49:42 INFO Client:54 - Uploading resource file:/tmp/spark-a6309dd8-4b21-4d65-801f-0ff209f19c4e/__spark_libs__8377589047505483518.zip -> hdfs://mycluster/user/root/.sparkStaging/application_1555593997898_0001/__spark_libs__8377589047505483518.zip
11. 2019-04-1821:49:49 INFO Client:54 - Uploading resource file:/software/spark-2.3.1/examples/jars/spark-examples_2.11-2.3.1.jar -> hdfs://mycluster/user/root/.sparkStaging/
```

```
application_1555593997898_0001/spark-examples_2.11-2.3.1.jar
2019-04-1821:49:50 INFO Client:54 - Uploading resource file:/tmp/spark-a6309dd8-4b21-4d65-801f-0ff209f19c4e/__spark_conf__6444994215348625691.zip -> hdfs://mycluster/user/root/.sparkStaging/application_1555593997898_0001/__spark_conf__.zip
2019-04-1821:49:50 INFO SecurityManager:54 - Changing view acls to: root
2019-04-1821:49:50 INFO SecurityManager:54 - Changing modify acls to: root
2019-04-1821:49:50 INFO SecurityManager:54 - Changing view acls groups to:
2019-04-1821:49:50 INFO SecurityManager:54 - Changing modify acls groups to:
2019-04-1821:49:50 INFO SecurityManager:54 - SecurityManager: authentication disabled; uiacls disabled; users with view permissions:Set(root); groups with view permissions:Set(); userswith modify permissions:Set(root); groups with modify permissions:Set()
2019-04-1821:49:50 INFO Client:54 - Submitting application application_1555593997898_0001 to ResourceManager
2019-04-1821:49:50 INFO YarnClientImpl:251 - Submitted application application_1555593997898_0001
2019-04-1821:49:51 INFO Client:54 - Application report for application_1555593997898_0001 (state: ACCEPTED)
2019-04-1821:49:51 INFO Client:54 -
client token: N/A
diagnostics: N/A
ApplicationMaster host: N/A
ApplicationMaster RPC port: -1
queue:default
start time:1555595390324
final status: UNDEFINED
tracking URL: http://mynode1:8088/proxy/application_1555593997898_0001/
user: root
2019-04-1821:49:52 INFO Client:54 - Application report for application_1555593997898_0001 (state: ACCEPTED)
```

```
32.2019 - 04 - 1821:49:53 INFO Client:54 - Application report for
   application_1555593997898_0001 (state: ACCEPTED)
33.2019 - 04 - 1821:49:54 INFO Client:54 - Application report for
   application_1555593997898_0001 (state: ACCEPTED)
34.2019 - 04 - 1821:49:55 INFO Client:54 - Application report for
   application_1555593997898_0001 (state: ACCEPTED)
35.2019 - 04 - 1821:49:56 INFO Client:54 - Application report for
   application_1555593997898_0001 (state: ACCEPTED)
36.2019 - 04 - 1821:49:57 INFO Client:54 - Application report for
   application_1555593997898_0001 (state: ACCEPTED)
37.2019 - 04 - 1821:49:58 INFO Client:54 - Application report for
   application_1555593997898_0001 (state: ACCEPTED)
38.2019 - 04 - 1821:49:59 INFO Client:54 - Application report for
   application_1555593997898_0001 (state: ACCEPTED)
39.2019 - 04 - 1821:50:00 INFO Client:54 - Application report for
   application_1555593997898_0001 (state: ACCEPTED)
40.2019 - 04 - 1821:50:01 INFO Client:54 - Application report for
   application_1555593997898_0001 (state: ACCEPTED)
41.2019 - 04 - 1821:50:02 INFO Client:54 - Application report for
   application_1555593997898_0001 (state: ACCEPTED)
42.2019 - 04 - 1821:50:03 INFO Client:54 - Application report for
   application_1555593997898_0001 (state: ACCEPTED)
43.2019 - 04 - 1821:50:04 INFO Client:54 - Application report for
   application_1555593997898_0001 (state: ACCEPTED)
44.2019 - 04 - 1821:50:05 INFO Client:54 - Application report for
   application_1555593997898_0001 (state: ACCEPTED)
45.2019 - 04 - 1821:50:06 INFO Client:54 - Application report for
   application_1555593997898_0001 (state: ACCEPTED)
46.2019 - 04 - 1821:50:07 INFO Client:54 - Application report for
   application_1555593997898_0001 (state: ACCEPTED)
47.2019 - 04 - 1821:50:08 INFO Client:54 - Application report for
   application_1555593997898_0001 (state: ACCEPTED)
48.2019 - 04 - 1821:50:09 INFO Client:54 - Application report for
   application_1555593997898_0001 (state: ACCEPTED)
49.2019 - 04 - 1821:50:10 INFO Client:54 - Application report for
   application_1555593997898_0001 (state: ACCEPTED)
50.2019 - 04 - 1821:50:11 INFO Client:54 - Application report for
   application_1555593997898_0001 (state: RUNNING)
```

```
2019-04-1821:50:11 INFO Client:54 -
client token: N/A
diagnostics: N/A
ApplicationMaster host:192.168.179.17
ApplicationMaster RPC port:0
queue:default
start time:1555595390324
final status: UNDEFINED
tracking URL: http://mynode1:8088/proxy/application_15555
93997898_0001/
user: root
2019-04-1821:50:12 INFO Client:54 - Application report for
application_1555593997898_0001 (state: RUNNING)
2019-04-1821:50:13 INFO Client:54 - Application report for
application_1555593997898_0001 (state: RUNNING)
2019-04-1821:50:14 INFO Client:54 - Application report for
application_1555593997898_0001 (state: RUNNING)
2019-04-1821:50:15 INFO Client:54 - Application report for
application_1555593997898_0001 (state: RUNNING)
2019-04-1821:50:16 INFO Client:54 - Application report for
application_1555593997898_0001 (state: RUNNING)
2019-04-1821:50:17 INFO Client:54 - Application report for
application_1555593997898_0001 (state: RUNNING)
2019-04-1821:50:18 INFO Client:54 - Application report for
application_1555593997898_0001 (state: RUNNING)
2019-04-1821:50:19 INFO Client:54 - Application report for
application_1555593997898_0001 (state: RUNNING)
2019-04-1821:50:20 INFO Client:54 - Application report for
application_1555593997898_0001 (state: RUNNING)
2019-04-1821:50:21 INFO Client:54 - Application report for
application_1555593997898_0001 (state: RUNNING)
2019-04-1821:50:22 INFO Client:54 - Application report for
application_1555593997898_0001 (state: RUNNING)
2019-04-1821:50:23 INFO Client:54 - Application report for
application_1555593997898_0001 (state: RUNNING)
2019-04-1821:50:24 INFO Client:54 - Application report for
application_1555593997898_0001 (state: RUNNING)
```

```
2019 - 04 - 1821:50:25 INFO Client:54 - Application report for application_1555593997898_0001 (state: FINISHED)
2019 -04 -1821:50:25 INFO Client:54 -
client token: N/A
diagnostics: N/A
ApplicationMaster host:192.168.179.17
ApplicationMaster RPC port:0
queue:default
start time:1555595390324
final status: SUCCEEDED
tracking URL: http://mynode1:8088/proxy/application_1555593997898_0001/
user: root
2019 - 04 - 1821:50:25 INFO ShutdownHookManager:54 - Shutdown hook called
2019 - 04 - 1821:50:25 INFO ShutdownHookManager:54 - Deleting directory /tmp/spark - a6309dd8 - 4b21 - 4d65 - 801f - 0ff209f19c4e
2019 - 04 - 1821:50:25 INFO ShutdownHookManager:54 - Deleting directory /tmp/spark - 0024eb48 - 0b3a - 4207 - b6d0 - d0d01e70f6cc
```

使用命令查看提交集群的任务日志是否保存到 HDFS 路径/spark/log 下：

```
[root@mynode4 bin]#hdfsdfs -ls /spark/log
19/04/1822:01:18 WARN util.NativeCodeLoader: Unable to load native - hadoop library for your platform...usingbuiltin - java classes where applicable
Found1 items
-rwxrwx---3 root supergroup 125542019 -04 -1821:50 /spark/log/application_1555593997898_0001_1.lz4
```

经过以上查询，发现当前 Spark 任务的日志已经被保存到 HDFS 路径中。当上述 Spark 任务运行完成之后，在浏览器中输入“https://mynode1:8088”，进入 Yarn 任务管理界面，检查已完成 Spark 任务的日志信息能否被成功查看。

单击图 4 -4 中“Tracking UI”对应的“History”，弹出图 4 -5 所示的页面，表示集群已经成功将已运行完的 Spark 任务日志还原。

单击 Spark 任务界面的“Executor”选项，查看 Spark 任务运行的 Executor 日志信息，如图 4 -6 所示。

单击图 4 -6 中“Logs”对应的信息，查看 Spark 任务运行时的一些日志信息。例如，

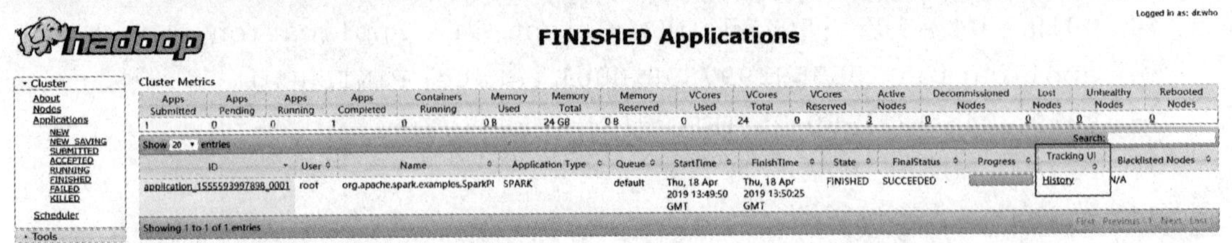


图 4-4　已完成任务列表

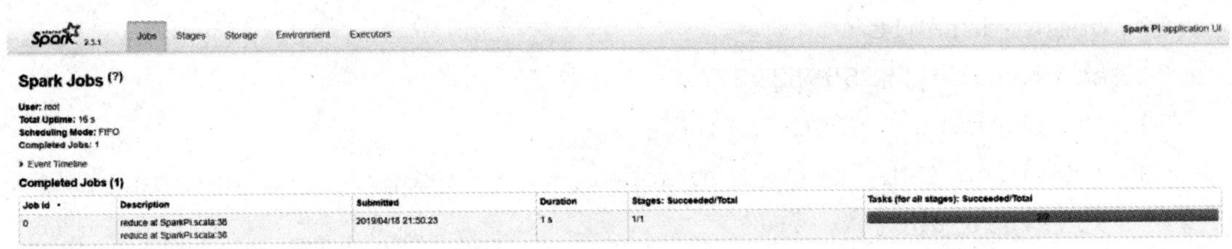


图 4-5　Spark 任务详情

Executors

Show 20 entries　　　　Search:

| Executor ID | Address | Status | RDD Blocks | Storage Memory | Disk Used | Cores | Active Tasks | Failed Tasks | Complete Tasks | Total Tasks | Task Time (GC Time) | Input | Shuffle Read | Shuffle Write | Logs |
|---|---|---|---|---|---|---|---|---|---|---|---|---|---|---|---|
| driver | mynode5:34093 | Active | 0 | 0.0 B / 384.1 MB | 0.0 B | 0 | 0 | 0 | 0 | 0 | 0 ms (0 ms) | 0.0 B | 0.0 B | 0.0 B | stdout stderr |
| 1 | mynode5:52692 | Active | 0 | 0.0 B / 384.1 MB | 0.0 B | 1 | 0 | 0 | 1 | 1 | 0.6 s (0.1 s) | 0.0 B | 0.0 B | 0.0 B | stdout stderr |
| 2 | mynode3:37210 | Active | 0 | 0.0 B / 384.1 MB | 0.0 B | 1 | 0 | 0 | 1 | 1 | 0.7 s (0.2 s) | 0.0 B | 0.0 B | 0.0 B | stdout stderr |

Showing 1 to 3 of 3 entries　　　　Previous 1 Next

图 4-6　Executor 信息

单击“Driver”，对应的标准输出日志信息如下：

```
LogType:stdout

LogUploadTime:18-四月-201921:50:30

LogLength:17302

2019-04-1821:49:59 INFO SignalUtils:54 - Registered signal han-
dler for TERM
2019-04-1821:49:59 INFO SignalUtils:54 - Registered signal han-
dler for HUP
2019-04-1821:49:59 INFO SignalUtils:54 - Registered signal han-
dler for INT
2019-04-1821:49:59 INFO SecurityManager:54 - Changing view
acls to: root
2019-04-1821:49:59 INFO SecurityManager:54 - Changing modify
acls to: root
```

```
12.2019-04-1821:49:59 INFO SecurityManager:54 - Changing view
   acls groups to:
13.2019-04-1821:49:59 INFO SecurityManager:54 - Changing modify
   acls groups to:
14.2019-04-1821:49:59 INFO SecurityManager:54 - SecurityManager:
   authentication disabled; uiacls disabled; users with view per-
   missions:Set(root); groups with view permissions:Set(); users
   with modify permissions:Set(root); groups with modify permis-
   sions:Set()
15.2019-04-1821:49:59 WARN NativeCodeLoader:62 - Unable to load
   native-hadoop library for your platform...usingbuiltin-java
   classes where applicable
16.2019-04-1821:49:59 INFO ApplicationMaster:54 - PreparingLocal
   resources
17.2019-04-1821:50:09 INFO ApplicationMaster:54 - ApplicationAt-
   temptId: appattempt_1555593997898_0001_000001
18.2019-04-1821:50:09 INFO ApplicationMaster:54 - Starting the
   user application in a separate Thread
19.2019-04-1821:50:09 INFO ApplicationMaster:54 - Waitingfor
   spark context initialization...
20.2019-04-1821:50:09 INFO SparkContext:54 - RunningSpark ver-
   sion 2.3.1
21.2019-04-1821:50:09 INFO SparkContext:54 - Submitted applica-
   tion:SparkPi
22.2019-04-1821:50:09 INFO SecurityManager:54 - Changing view
   acls to: root
23.2019-04-1821:50:09 INFO SecurityManager:54 - Changing modify
   acls to: root
24.2019-04-1821:50:09 INFO SecurityManager:54 - Changing view
   acls groups to:
25.2019-04-1821:50:09 INFO SecurityManager:54 - Changing modify
   acls groups to:
26.2019-04-1821:50:09 INFO SecurityManager:54 - SecurityManager:
   authentication disabled; uiacls disabled; users with view per-
   missions:Set(root); groups with view permissions:Set(); users
   with modify permissions:Set(root); groups with modify permis-
   sions:Set()
```

```
27. 2019-04-1821:50:09 INFO Utils:54 - Successfully started serv-
    ice 'sparkDriver' on port 50738.
28. 2019-04-1821:50:09 INFO SparkEnv:54 - RegisteringMapOutput-
    Tracker
29. 2019-04-1821:50:09 INFO SparkEnv:54 - RegisteringBlockMan-
    agerMaster
30. 2019-04-1821:50:09 INFO BlockManagerMasterEndpoint:54 - Usin-
    gorg.apache.spark.storage.DefaultTopologyMapperfor getting
    topology information
31. 2019-04-1821:50:09 INFO BlockManagerMasterEndpoint:54 -
    BlockManagerMasterEndpoint up
32. 2019-04-1821:50:09 INFO DiskBlockManager:54 - Createdlocal di-
    rectory at /opt/data/hadoop/nm-local-dir/usercache/root/
    appcache/application_1555593997898_0001/blockmgr-7e4c99ef-
    e7bb-439e-a3ad-f5c0782e7aeb
33. 2019-04-1821:50:09 INFO MemoryStore:54 - MemoryStore started
    with capacity 366.3 MB
34. 2019-04-1821:50:09 INFO SparkEnv:54 - RegisteringOutputCom-
    mitCoordinator
35. 2019-04-1821:50:09 INFO log:192 - Logging initialized
    @11461ms
36. 2019-04-1821:50:09 INFO JettyUtils:54 - Adding filter:org.
    apache.hadoop.yarn.server.webproxy.amfilter.AmIpFilter
37. 2019-04-1821:50:09 INFO Server:346 - jetty-9.3.z-SNAPSHOT
38. 2019-04-1821:50:10 INFO Server:414 - Started@11545ms
39. 2019-04-1821:50:10 INFO AbstractConnector:278 - Started-
    ServerConnector@762b1145{HTTP/1.1,[http/1.1]}{0.0.0.0:53431}
40. 2019-04-1821:50:10 INFO Utils:54 - Successfully started serv-
    ice 'SparkUI' on port 53431.
41. 2019-04-1821:50:10 INFO ContextHandler:781 - Started o.s.j.s.
    ServletContextHandler@5aba0368{/jobs,null,AVAILABLE,@Spark}
42. 2019-04-1821:50:10 INFO ContextHandler:781 - Started o.s.j.s.
    ServletContextHandler@5d0e30a6{/jobs/json,null,AVAILABLE,
    @Spark}
43. 2019-04-1821:50:10 INFO ContextHandler:781 - Started o.s.j.s.
    ServletContextHandler@8fb8e22{/jobs/job,null,AVAILABLE,
    @Spark}
```

```
44.2019-04-18 21:50:10 INFO ContextHandler:781 - Started o.s.j.s.
   ServletContextHandler@199027e{/jobs/job/json,null,AVAILA-
   BLE,@Spark}
45.2019-04-18 21:50:10 INFO ContextHandler:781 - Started o.s.j.s.
   ServletContextHandler@2d3e89eb{/stages,null,AVAILABLE,
   @Spark}
46.2019-04-18 21:50:10 INFO ContextHandler:781 - Started o.s.j.s.
   ServletContextHandler@79e098f0{/stages/json,null,AVAILABLE,
   @Spark}
47.2019-04-18 21:50:10 INFO ContextHandler:781 - Started o.s.j.s.
   ServletContextHandler@16fb7485{/stages/stage,null,AVAILA-
   BLE,@Spark}
48.2019-04-18 21:50:10 INFO ContextHandler:781 - Started o.s.j.s.
   ServletContextHandler@7b4a9a31{/stages/stage/json,null,
   AVAILABLE,@Spark}
49.2019-04-18 21:50:10 INFO ContextHandler:781 - Started o.s.j.s.
   ServletContextHandler@36588681{/stages/pool,null,AVAILABLE,
   @Spark}
50.2019-04-18 21:50:10 INFO ContextHandler:781 - Started o.s.j.s.
   ServletContextHandler@29897032{/stages/pool/json,null,
   AVAILABLE,@Spark}
51.2019-04-18 21:50:10 INFO ContextHandler:781 - Started o.s.j.s.
   ServletContextHandler@2a32a075{/storage,null,AVAILABLE,
   @Spark}
52.2019-04-18 21:50:10 INFO ContextHandler:781 - Started o.s.j.s.
   ServletContextHandler@5062c10{/storage/json,null,AVAILABLE,
   @Spark}
53.2019-04-18 21:50:10 INFO ContextHandler:781 - Started o.s.j.s.
   ServletContextHandler@7718debf{/storage/rdd,null,AVAILABLE,
   @Spark}
54.2019-04-18 21:50:10 INFO ContextHandler:781 - Started o.s.j.s.
   ServletContextHandler@1572288e{/storage/rdd/json,null,
   AVAILABLE,@Spark}
55.2019-04-18 21:50:10 INFO ContextHandler:781 - Started o.s.j.s.
   ServletContextHandler@2afc035d{/environment,null,AVAILABLE,
   @Spark}
56.2019-04-18 21:50:10 INFO ContextHandler:781 - Started o.s.j.s.
```

```
    ServletContextHandler @ 1e2eccbe {/ environment/ json, null,
    AVAILABLE,@Spark}
57. 2019-04-18 21:50:10 INFO ContextHandler:781 - Started o.s.j.s.
    ServletContextHandler@2ff214e2{/executors,null,AVAILABLE,
    @Spark}
58. 2019-04-18 21:50:10 INFO ContextHandler:781 - Started o.s.j.s.
    ServletContextHandler@2bc06fca{/executors/json,null,AVAILA-
    BLE,@Spark}
59. 2019-04-18 21:50:10 INFO ContextHandler:781 - Started o.s.j.s.
    ServletContextHandler@61ec7704{/executors/threadDump,null,
    AVAILABLE,@Spark}
60. 2019-04-18 21:50:10 INFO ContextHandler:781 - Started o.s.j.s.
    ServletContextHandler@94f0cbb{/executors/threadDump/json,
    null,AVAILABLE,@Spark}
61. 2019-04-18 21:50:10 INFO ContextHandler:781 - Started o.s.j.s.
    ServletContextHandler @ 36254c74 {/ static, null, AVAILABLE,
    @Spark}
62. 2019-04-18 21:50:10 INFO ContextHandler:781 - Started o.s.j.s.
    ServletContextHandler@9d046e9{/,null,AVAILABLE,@Spark}
63. 2019-04-18 21:50:10 INFO ContextHandler:781 - Started o.s.j.s.
    ServletContextHandler@24b24737{/api,null,AVAILABLE,@Spark}
64. 2019-04-18 21:50:10 INFO ContextHandler:781 - Started o.s.j.s.
    ServletContextHandler@31b27117{/jobs/job/kill,null,AVAILA-
    BLE,@Spark}
65. 2019-04-18 21:50:10 INFO ContextHandler:781 - Started o.s.j.s.
    ServletContextHandler @ 55e8bbd2 {/ stages/ stage/ kill, null,
    AVAILABLE,@Spark}
66. 2019-04-18 21:50:10 INFO SparkUI:54 - BoundSparkUI to 0.0.0.0,
    and started at http://mynode5:53431
67. 2019-04-18 21:50:10 INFO YarnClusterScheduler:54 - Creat-
    edYarnClusterScheduler
68. 2019-04-18 21:50:10 INFO SchedulerExtensionServices:54 - Star-
    tingYarn extension services with app application_1555593997898_
    0001 andattemptIdSome(appattempt_1555593997898_0001_000001)
69. 2019-04-18 21:50:10 INFO Utils:54 - Successfully started service
    'org.apache.spark.network.netty.NettyBlockTransferService' on
    port 34093.
```

```
2019-04-18 21:50:10 INFO NettyBlockTransferService:54 - Server created on mynode5:34093
2019-04-18 21:50:10 INFO BlockManager:54 - Using org.apache.spark.storage.RandomBlockReplicationPolicy for block replication policy
2019-04-18 21:50:10 INFO BlockManagerMaster:54 - Registering BlockManager BlockManagerId(driver, mynode5,34093,None)
2019-04-18 21:50:10 INFO BlockManagerMasterEndpoint:54 - Registering block manager mynode5:34093 with 366.3 MB RAM, BlockManagerId(driver, mynode5,34093,None)
2019-04-18 21:50:10 INFO BlockManagerMaster:54 - Registered BlockManager BlockManagerId(driver, mynode5,34093,None)
2019-04-18 21:50:10 INFO BlockManager:54 - Initialized BlockManager: BlockManagerId(driver, mynode5,34093,None)
2019-04-18 21:50:10 INFO ContextHandler:781 - Started o.s.j.s.ServletContextHandler@414ca774{/metrics/json,null,AVAILABLE,@Spark}
2019-04-18 21:50:10 INFO EventLoggingListener:54 - Logging events to hdfs://mycluster/spark/log/application_1555593997898_0001_1.lz4
2019-04-18 21:50:10 INFO ApplicationMaster:54 -
===============================================================================
YARN executor launch context:
env:
CLASSPATH -> {{PWD}}<CPS>{{PWD}}/__spark_conf__<CPS>{{PWD}}/__spark_libs__/*<CPS>$HADOOP_CONF_DIR<CPS>$HADOOP_COMMON_HOME/share/hadoop/common/*<CPS>$HADOOP_COMMON_HOME/share/hadoop/common/lib/*<CPS>$HADOOP_HDFS_HOME/share/hadoop/hdfs/*<CPS>$HADOOP_HDFS_HOME/share/hadoop/hdfs/lib/*<CPS>$HADOOP_YARN_HOME/share/hadoop/yarn/*<CPS>$HADOOP_YARN_HOME/share/hadoop/yarn/lib/*<CPS>$HADOOP_MAPRED_HOME/share/hadoop/mapreduce/*<CPS>$HADOOP_MAPRED_HOME/share/hadoop/mapreduce/lib/*<CPS>{{PWD}}/__spark_conf__/__hadoop_conf__
SPARK_YARN_STAGING_DIR -> hdfs://mycluster/user/root/.sparkStaging/application_1555593997898_0001
```

```
SPARK_USER -> root

command:
{{JAVA_HOME}}/bin/java \
-server \
-Xmx1024m \
-Djava.io.tmpdir={{PWD}}/tmp \
-Dspark.yarn.app.container.log.dir=<LOG_DIR> \
-XX:OnOutOfMemoryError='kill %p' \
org.apache.spark.executor.CoarseGrainedExecutorBackend \
--driver-url \
spark://CoarseGrainedScheduler@mynode5:50738 \
--executor-id \
<executorId> \
--hostname \
<hostname> \
--cores \
1 \
--app-id \
application_1555593997898_0001 \
--user-class-path \
file:$PWD/__app__.jar \
1><LOG_DIR>/stdout \
2><LOG_DIR>/stderr

resources:
__app__.jar -> resource { scheme: "hdfs" host: "mycluster" port: -1 file: "/user/root/.sparkStaging/application_1555593997898_0001/spark-examples_2.11-2.3.1.jar" } size: 1997556 timestamp: 1555595389918 type: FILE visibility: PRIVATE
__spark_libs__ -> resource { scheme: "hdfs" host: "mycluster" port: -1 file: "/user/root/.sparkStaging/application_1555593997898_0001/__spark_libs__8377589047505483518.zip" } size: 232213520 timestamp: 1555595387662 type: ARCHIVE visibility: PRIVATE
__spark_conf__ -> resource { scheme: "hdfs" host: "mycluster" port: -1 file: "/user/root/.sparkStaging/application_
```

```
    1555593997898_0001/__spark_conf__.zip"} size:220559 times-
    tamp:1555595390167 type:ARCHIVE visibility:PRIVATE
113.
114.=====================================================================
    =================
115.2019-04-18 21:50:11 INFO YarnRMClient:54 - Registering
    the ApplicationMaster
116.2019-04-18 21:50:11 INFO YarnSchedulerBackend$YarnSchedul-
    erEndpoint:54 - ApplicationMaster registered as NettyRpcEnd-
    pointRef(spark://YarnAM@mynode5:50738)
117.2019-04-18 21:50:11 INFO YarnAllocator:54 - Will request 2
    executor container(s), each with 1 core(s) and 1408 MB memory
    (including 384 MB of overhead)
118.2019-04-18 21:50:11 INFO YarnAllocator:54 - Submitted 2 unlo-
    calized container requests.
119.2019-04-18 21:50:11 INFO ApplicationMaster:54 - Started pro-
    gress reporter thread with (heartbeat : 3000, initial alloca-
    tion : 200) intervals
120.2019-04-18 21:50:12 INFO AMRMClientImpl:361 - Received new
    token for : mynode5:45911
121.2019-04-18 21:50:12 INFO AMRMClientImpl:361 - Received new
    token for : mynode3:45349
122.2019-04-18 21:50:12 INFO YarnAllocator:54 - Launching con-
    tainer container_1555593997898_0001_01_000002 on host
    mynode5 for executor with ID 1
123.2019-04-18 21:50:12 INFO YarnAllocator:54 - Launching con-
    tainer container_1555593997898_0001_01_000003 on host my-
    node3 for executor with ID 2
124.2019-04-18 21:50:12 INFO YarnAllocator:54 - Received 2 con-
    tainers from YARN, launching executors on 2 of them.
125.2019-04-18 21:50:12 INFO ContainerManagementProtocolProxy:
    81 - yarn.client.max-cached-nodemanagers-proxies : 0
126.2019-04-18 21:50:12 INFO ContainerManagementProtocolProxy:
    81 - yarn.client.max-cached-nodemanagers-proxies : 0
127.2019-04-18 21:50:12 INFO ContainerManagementProtocolProxy:
    260 - Opening proxy : mynode5:45911
128.2019-04-18 21:50:12 INFO ContainerManagementProtocolProxy:
    260 - Opening proxy : mynode3:45349
```

```
129.2019-04-18 21:50:14 INFO YarnSchedulerBackend$YarnDriver-
    Endpoint:54 - Registered executor NettyRpcEndpointRef(spark -
    client://Executor) (192.168.179.17:38759) with ID 1
130.2019-04-18 21:50:15 INFO BlockManagerMasterEndpoint:54 - Reg-
    istering block manager mynode5:52692 with 366.3 MB RAM, Block-
    ManagerId(1, mynode5, 52692, None)
131.2019-04-18 21:50:23 INFO YarnSchedulerBackend$YarnDriver-
    Endpoint:54 - Registered executor NettyRpcEndpointRef(spark -
    client://Executor) (192.168.179.15:33254) with ID 2
132.2019-04-18 21:50:23 INFO YarnClusterSchedulerBackend:54 -
    SchedulerBackend is ready for scheduling beginning after
    reached minRegisteredResourcesRatio: 0.8
133.2019-04-18 21:50:23 INFO YarnClusterScheduler:54 - YarnClus-
    terScheduler.postStartHook done
134.2019-04-18 21:50:23 INFO BlockManagerMasterEndpoint:54 - Reg-
    istering block manager mynode3:37210 with 366.3 MB RAM, Block-
    ManagerId(2, mynode3, 37210, None)
135.2019-04-18 21:50:23 INFO SparkContext:54 - Starting job: re-
    duce at SparkPi.scala:38
136.2019-04-18 21:50:23 INFO DAGScheduler:54 - Got job 0 (reduce
    at SparkPi.scala:38) with 2 output partitions
137.2019-04-18 21:50:23 INFO DAGScheduler:54 - Final stage: Re-
    sultStage 0 (reduce at SparkPi.scala:38)
138.2019-04-18 21:50:23 INFO DAGScheduler:54 - Parents of final
    stage: List()
139.2019-04-18 21:50:23 INFO DAGScheduler:54 - Missing parents:
    List()
140.2019-04-18 21:50:23 INFO DAGScheduler:54 - Submitting Result-
    Stage 0 (MapPartitionsRDD[1] at map at SparkPi.scala:34),
    which has no missing parents
141.2019-04-18 21:50:23 INFO MemoryStore:54 - Block broadcast_0 stored
    as values in memory (estimated size 1832.0 B, free 366.3 MB)
142.2019-04-18 21:50:24 INFO MemoryStore:54 - Block broadcast_0_
    piece0 stored as bytes in memory (estimated size 1181.0 B, free
    366.3 MB)
143.2019-04-18 21:50:24 INFO BlockManagerInfo:54 - Added broadcast_0_
    piece0 in memory on mynode5:34093 (size: 1181.0 B, free: 366.3 MB)
```

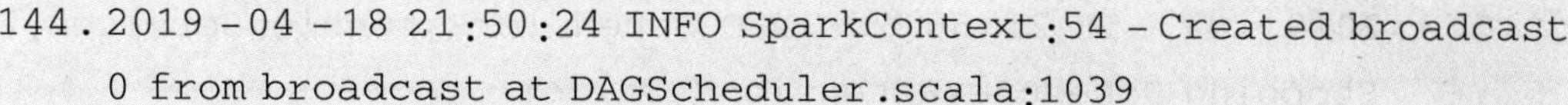



```
144.2019-04-18 21:50:24 INFO SparkContext:54 - Created broadcast
    0 from broadcast at DAGScheduler.scala:1039
145.2019-04-18 21:50:24 INFO DAGScheduler:54 - Submitting 2 missing
    tasks from ResultStage 0 (MapPartitionsRDD[1] at map at SparkPi.
    scala:34) (first 15 tasks are for partitions Vector(0, 1))
146.2019-04-18 21:50:24 INFO YarnClusterScheduler:54 - Adding
    task set 0.0 with 2 tasks
147.2019-04-18 21:50:24 INFO TaskSetManager:54 - Starting task 0.0
    in stage 0.0 (TID 0, mynode3, executor 2, partition 0, PROCESS_LO-
    CAL, 7842 bytes)
148.2019-04-18 21:50:24 INFO TaskSetManager:54 - Starting task
    1.0 in stage 0.0 (TID 1, mynode5, executor 1, partition 1,
    PROCESS_LOCAL, 7842 bytes)
149.2019-04-18 21:50:24 INFO BlockManagerInfo:54 - Added broadcast_0_
    piece0 in memory on mynode5:52692 (size: 1181.0 B, free: 366.3 MB)
150.2019-04-18 21:50:24 INFO BlockManagerInfo:54 - Added broadcast_0_
    piece0 in memory on mynode3:37210 (size: 1181.0 B, free: 366.3 MB)
151.2019-04-18 21:50:24 INFO TaskSetManager:54 - Finished task
    1.0 in stage 0.0 (TID 1) in 602 ms on mynode5 (executor 1) (1/2)
152.2019-04-18 21:50:24 INFO TaskSetManager:54 - Finished task
    0.0 in stage 0.0 (TID 0) in 696 ms on mynode3 (executor 2) (2/2)
153.2019-04-18 21:50:24 INFO YarnClusterScheduler:54 - Removed
    TaskSet 0.0, whose tasks have all completed, from pool
154.2019-04-18 21:50:24 INFO DAGScheduler:54 - ResultStage 0 (re-
    duce at SparkPi.scala:38) finished in 1.188 s
155.2019-04-18 21:50:24 INFO DAGScheduler:54 - Job 0 finished:
    reduce at SparkPi.scala:38, took 1.256346 s
156.Pi is roughly 3.144275721378607
157.2019-04-18 21:50:24 INFO AbstractConnector:318 - Stopped
    Spark@762b1145{HTTP/1.1,[http/1.1]}{0.0.0.0:0}
158.2019-04-18 21:50:24 INFO SparkUI:54 - Stopped Spark web UI at
    http://mynode5:53431
159.2019-04-18 21:50:25 INFO YarnAllocator:54 - Driver requested
    a total number of 0 executor(s).
160.2019-04-18 21:50:25 INFO YarnClusterSchedulerBackend:54 -
    Shutting down all executors
161.2019-04-18 21:50:25 INFO YarnSchedulerBackend$YarnDriver-
    Endpoint:54 - Asking each executor to shut down
```

```
2019 - 04 - 18 21:50:25 INFO SchedulerExtensionServices:54 - Stopping SchedulerExtensionServices
(serviceOption = None,
services = List(),
started = false)
2019 - 04 - 18 21:50:25 INFO MapOutputTrackerMasterEndpoint:54 - MapOutputTrackerMasterEndpoint stopped!
2019 -04 -18 21:50:25 INFO MemoryStore:54 - MemoryStore cleared
2019 -04 -18 21:50:25 INFO BlockManager:54 - BlockManager stopped
2019 - 04 - 18 21:50:25 INFO BlockManagerMaster:54 - BlockManagerMaster stopped
2019 -04 -18 21:50:25 INFO OutputCommitCoordinator$OutputCommitCoordinatorEndpoint:54 - OutputCommitCoordinator stopped!
2019 - 04 - 18 21:50:25 INFO SparkContext: 54 - Successfully stopped SparkContext
2019 - 04 - 18 21:50:25 INFO ApplicationMaster:54 - Final app status: SUCCEEDED, exitCode: 0
2019 -04 -18 21:50:25 INFO ApplicationMaster:54 - Unregistering ApplicationMaster with SUCCEEDED
2019 -04 -18 21:50:25 INFO AMRMClientImpl:383 - Waiting for application to be successfully unregistered.
2019 - 04 - 18 21:50:25 INFO ApplicationMaster:54 - Deleting staging directory hdfs://mycluster/user/root/.sparkStaging/application_1555593997898_0001
2019 - 04 - 18 21:50:25 INFO ShutdownHookManager:54 - Shutdown hook called
2019 - 04 - 18 21:50:25 INFO ShutdownHookManager:54 - Deleting directory /opt/data/hadoop/nm - local - dir/usercache/root/appcache/application_1555593997898_0001/spark - 765467ed - 2ae5 -4cc3 -9ef7 -6e7f85fc40a2
```

此时，查看 Spark 日志信息的操作完成。可以通过以上方式在 Yarn 的任务管理界面中查看提交到 Yarn 集群的 Spark 任务日志信息。

## 任务 4.2　验证日志服务器

为了巩固以上 Spark 历史日志的配置使用，以 Spark - Shell 启动为例，验证 Spark 任务的日志信息能否保存到 Spark 日志服务器。

### 4.2.1　Spark – Shell 介绍

Spark – Shell 是 Spark 自带的一个快速原型开发工具，也可以说是 Spark 的 Scala REPL（Read – Eval – Print – Loop），即交互式 Shell。支持使用 Scala 语言进行 Spark 的交互式编程。

可以在提交 Spark 任务的客户端 mynode4 上启动 Spark – Shell，命令如下：

```
[root@mynode4 bin]# cd /software/spark-2.3.1/bin/
[root@mynode4 bin]#./spark-shell --master yarn --deploy-mode cluster
Error:Cluster deploy mode isnot applicable to Spark shells.
Runwith --help for usage help or --verbose for debug output
[root@mynode4 bin]#./spark-shell --master yarn
2019-04-1822:36:58 WARN NativeCodeLoader:62 - Unable to load native-hadoop library for your platform...usingbuiltin-java classes where applicable
Settingdefault log level to "WARN".
To adjust logging level usesc.setLogLevel(newLevel).ForSparkR, usesetLogLevel(newLevel).
2019-04-1822:37:09 WARN Client:66 - Neitherspark.yarn.jars nor spark.yarn.archiveisset, falling back to uploading libraries under SPARK_HOME.
Spark context Web UI available at http://mynode4:4040
Spark context available as 'sc'(master = yarn, app id = application_1555593997898_0002).
Spark session available as 'spark'.
Welcome to
____ __
/__/___ ____//__
_\ \/_ \/_'/__/'_/
/__/.__/\_,_/_//_/\_\version 2.3.1
/_/

Using Scala version 2.11.8 (Java HotSpot(TM) 64-Bit Server VM, Java 1.8.0_181)
Type in expressions to have them evaluated.
Type :help for more information.

scala >
```

### 4.2.2 查看运行日志

打开浏览器，输入“https://mynode1:8088”，进入 Yarn 任务管理页面，会看到处于运行状态的 Spark 任务就是当前提交的任务。可以通过单击页面上的“Tracking UI”进入任务运行时的页面，查看任务运行过程中的状态信息，如图 4-7 所示。

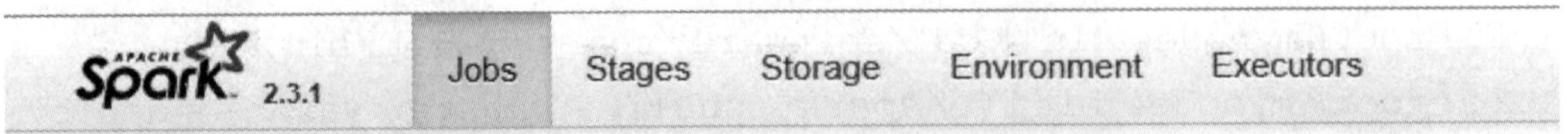


图 4-7　Spark 任务运行时页面

### 4.2.3 启动日志服务器

图 4-7 所示的页面查询信息是 Spark 任务运行时的状态信息，可以将 Spark-Shell 关闭，确认是否可以查看已完成任务的日志状态信息。要保存 Spark-Shell 的运行日志信息，需要启动 Spark 的历史日志服务器及 Yarn 的历史日志服务器，命令如下：

```
#在客户端上启动历史日志服务器
[root@mynode4 ~]# cd /software/spark-2.3.1/sbin/
[root@mynode4 sbin]# ./start-history-server.sh
starting org.apache.spark.deploy.history.HistoryServer, logging to /software/spark-2.3.1/logs/spark-root-org.apache.spark.deploy.history.HistoryServer-1-mynode4.out

#在 mynode1 节点上启动 Yarn 的历史日志服务器
[root@mynode1 ~]# mr-jobhistory-daemon.sh start historyserver
starting historyserver, logging to /software/hadoop-2.6.5/logs/mapred-root-historyserver-mynode1.out
```

经过以上启动历史日志服务器之后，可以在 mynode4 上使用快捷键 Ctrl+C 将 Spark-Shell 关闭，查看 Spark-Shell 的日志是否保存成功。

### 4.2.4 查看日志

进入 Yarn 的日志管理页面“https://mynode1:8088”，在已完成任务列表中找到 Spark-Shell 任务，单击“Tracking UI”对应的“History”，弹出 Spark 任务运行完成的界面，如图 4-8 所示。

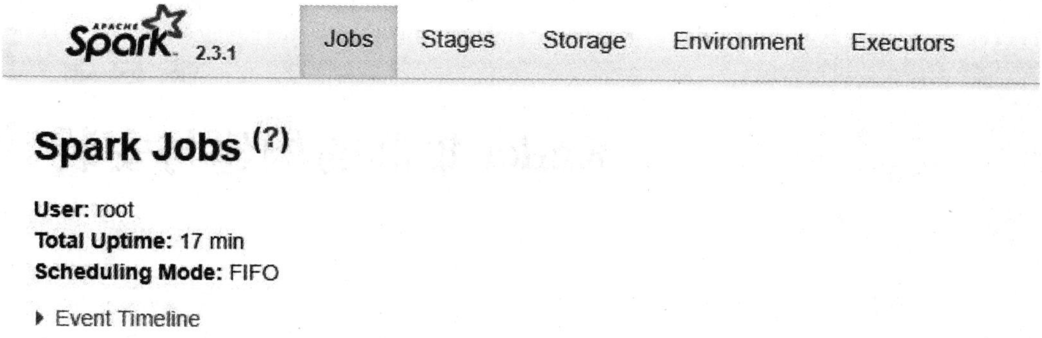

图 4－8　Spark 运行完成界面查看

经过上面的查询，可以看到已经运行成功的 Spark 任务的日志，通过 Spark 历史日志服务器还原回来，这样利于查看已完成任务的资源配置及日志报告信息，便于下次运行任务的参数调节。

# 项目五
# Kafka 集群的构建与安装

【项目描述】

在××系统中，数据来源于数据采集模块 Flume。经过 Flume 的数据需要使用 SparkStreaming 进行实时分析，以防 Flume 中的数据在某一时段过多而影响 SparkStreaming 稳定性。但是，如果将 Flume 的数据存入分布式数据库或者磁盘中，再使用 SparkStreaming 进行处理，由于分布式数据库和磁盘在读取和存储大量数据时有大量磁盘 I/O，效率低下，这样就背离了实时处理数据的初衷。

【项目分析】

为了保证数据源头的数据在峰值时可以被全量接收，同时，又能使 SparkStreaming 对流数据稳定处理，可以选择消息中间件 Kafka 来解决。数据首先接入 Kafka，然后 SparkStreaming 读取 Kafka 中的数据。Kafka 起到缓冲数据、解耦的作用。

## 任务 5.1　集群搭建

Kafka 是一个支持分区（partition）、多副本（replica）、基于 ZooKeeper 的协调的分布式消息系统，它的最大的特点就是可以实时地处理大量数据，以满足各种需求场景。比如基于 Hadoop 的批处理系统、低延迟的实时系统、Storm/Spark 流式处理引擎、Web/Nginx 日志、访问日志、消息服务等。

Kafka 的特点如下。

可扩展：Kafka 支持热扩展，可以动态增加节点及存储单元。

持久性、可靠性：消息不经过内存持久化到本地磁盘，并且支持数据备份，防止数据丢失。可以由用户来指定数据备份的份数，管理灵活。

容错性：允许集群中节点失败，会有其他节点代替当前节点进行数据传输。

高并发：Kafka 内部使用了零拷贝技术，就是数据在节点之间保存，传输时不需要经过

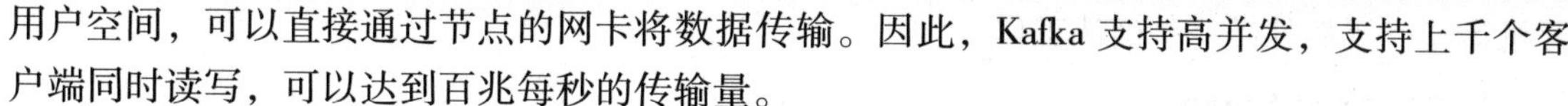


用户空间，可以直接通过节点的网卡将数据传输。因此，Kafka 支持高并发，支持上千个客户端同时读写，可以达到百兆每秒的传输量。

完整的 Kafka 集群节点划分应当包含以下两个部分：

**1. Broker**

Broker 是组成 Kafka 集群的节点，一个 Kafka 集群可以由多个 Broker 节点组成。Kafka 是消费订阅模式，Kafka 数据的源头叫作 Kafka 的生产者，消费 Kafka 的一端叫作 Kafka 的消费者。Broker 负责消息的读写请求，保存来自生产者的数据和为消费者提供消息，除此之外，Broker 还负责消息副本的保存。

**2. ZooKeeper**

ZooKeeper 是分布式协调系统。在 KafKa 集群中，ZooKeeper 负责保存 Kafka 集群中的元数据信息，这些元数据包括 Broker 节点信息、Topic 信息、Partition 信息及各种配置状态。在 Kafka 0. 8 版本之前，ZooKeeper 还负责保存管理消费者的消费消息位置（offset）。

### 5. 1. 1　运行平台支持

理论上，Kafka 可以运行在 Linux、Windows、MacOS、UNIX 等操作系统上。其中基于 Linux 安装是最常见的一种，因为 Linux 系统开源、免费且支持软件多，大多数服务器都选择 Linux 系统，很多数据处理技术也是基于 Liunx 处理。因此，在 Liunx 系统中安装 Kafka 是首要选择。

### 5. 1. 2　软件环境

Kafka 使用 Scala 语言编写，由于 Scala 是基于 JVM 运行的，所以，无论在哪个系统中搭建 Kafka 集群，节点上都要安装 JDK。

除 Java 以外，还需要在集群内的全部主机中都安装 SSH 并且保证 SSHD 进程一直运行。配置主机之间的免密登录能够更好地保证 Kafka 运行过程中不需要大量的人工介入，这就需要用到节点之间的免密通信手段。

安装 Kafka 还需要分布式协调系统 ZooKeeper。

### 5. 1. 3　集群构建流程

通常构建一个 Kafka 分布式集群需要完成如下工作步骤：

**1. 规划 Kafka 集群**

在安装之前，要根据硬件条件及项目需求对集群进行整体的规划，通常包括主机规划、软件规划、网络拓扑结构规划及集群规划。

**2. 构建运行平台**

按照集群规划方案进行安装主机操作系统、修改主机名称、配置主机 IP 地址等工作。

**3. 安装配置软件**

按照集群规划方案配置集群间的免密登录，为每台主机安装 JDK。为了确保 SSHD 能够开机运行，要额外做一些确认工作。

**4. 安装配置 Kafka**

按照集群规划方案为每台主机安装 Kafka，并将 Kafka 配置为与集群规划方案一致的角色。

**5. 启动 Kafka 并验证**

此时，Kafka 已经按照集群规划完成了全部的安装部署。为了确保安装配置的 Kafka 能够正常运行，要启动 Kafka 的各个进程，查看各节点的状态，查看 ZooKeeper 中的状态信息。

## 任务 5.2　集群规划

本任务将学习如何安装、配置、启动 Kafka 集群。首先对即将部署的 Kafka 集群进行总体规划。

规划过程中，使用 5 台服务器，默认这 5 台服务器已经分别修改好了主机名称，主机名称分别为 mynode1、mynode2、mynode3、mynode4、mynode5。其中，mynode1 ~ mynode3 集群节点是搭建 Kafka 集群使用的节点，mynode3 ~ mynode5 集群节点是搭建 ZooKeeper 分布式协调系统的节点。

### 5.2.1　集群节点划分

集群节点划分见表 5－1。

表 5－1　集群节点划分

| 主机名称 | IP 地址 | Kafka 角色名称 | ZooKeeper 角色名称 | CPU 核心数 | 内存 /GB | 硬盘容量 /GB |
|---|---|---|---|---|---|---|
| mynode1 | 192.168.179.13 | Broker | — | 8 核 64 位（不少于 4 核） | 20 | 500 以上 |
| mynode2 | 192.168.179.14 | Broker | — | 8 核 64 位（不少于双核） | 20 | 500 以上 |
| mynode3 | 192.168.179.15 | Broker | Leader 或者 Follower | 8 核 64 位（不少于双核） | 20 | 500 以上 |
| mynode4 | 192.168.179.16 | — | Leader 或者 Follower | 8 核 64 位（不少于双核） | 20 | 500 以上 |
| mynode5 | 192.168.179.17 | — | Leader 或者 Follower | 8 核 64 位（不少于双核） | 20 | 500 以上 |

### 5.2.2 软件选择

软件选择见表 5-2。

表 5-2 软件选择

| 软件 | 版本号 | 位数 | 版本说明 |
|---|---|---|---|
| 操作系统 | CentOS 6.5 | 64 位 | |
| JDK | 1.8.x | 64 位 | |
| Kafka | 0.11.x | — | 稳定版本 |
| ZooKeeper | 3.4.x | — | 稳定版本 |

注：以上 Kafka 版本是 0.11 版本。由于本书中使用的 Spark 是 2.3.1 版本，与 Spark 2.3.1 兼容的 Kafka 版本是 Kafka 0.10 及以上。这里选择 Kafka 0.11 版本。

除非特别说明，本书中所有的系统操作及命令均运行在 CentOS 6.5 操作系统上，其他 Linux 操作系统的命令与 CentOS 可能略有不同，请各位读者自行查阅相关资料。

### 5.2.3 网络结构规划

网络拓扑结构如图 5-1 所示。

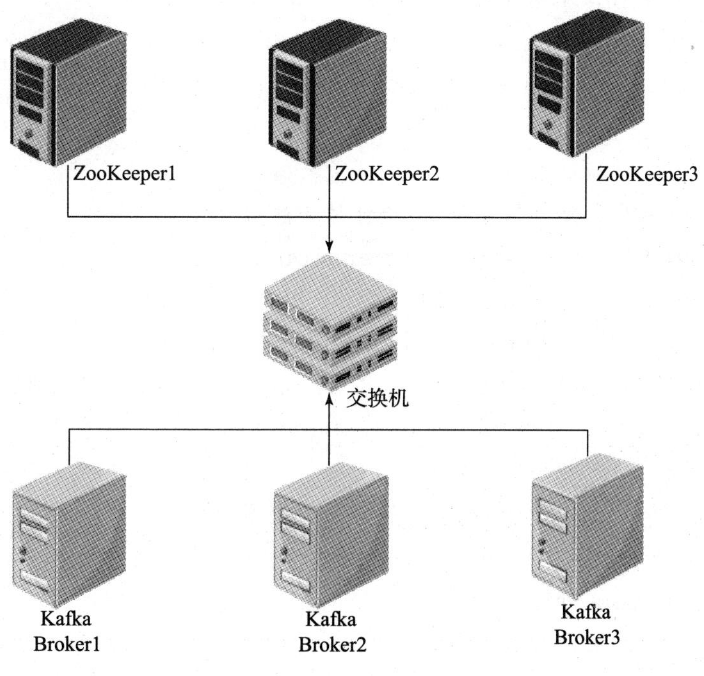


图 5-1 网络拓扑结构

## 任务 5.3　安装准备

安装 Kafka 之前，需要在 Kafka 集群每台节点上配置 JDK，节点之间免密。安装 JDK 是由于 Kafka 底层的源码是用 Scala 语言编写的，所以底层部署 Kafka 需要 JDK 环境。节点之间免密是由于 Kafka 集群 Broker 之间会有消息的传递复制。此外，还需要安装 ZooKeeper 集群，负责保存元数据及状态数据。

在介绍 Spark 安装时，已经为每台节点安装了 JDK8。这里的节点包含 mynode1、mynode2、mynode3、mynode4、mynode5，先验证集群中这五台节点的 JDK 安装配置信息。验证如下：

```
#在 mynode1 上执行如下命令,验证是否配置 JDK
[root@mynode1 ~]# java -version
java version "1.8.0_181"
Java(TM) SE Runtime Environment (build 1.8.0_181-b13)
Java HotSpot(TM) 64-Bit Server VM (build 25.181-b13, mixed mode)

#在 mynode2 上执行如下命令,验证是否配置 JDK
[root@mynode2 ~]# java -version
java version "1.8.0_181"
Java(TM) SE Runtime Environment (build 1.8.0_181-b13)
Java HotSpot(TM) 64-Bit Server VM (build 25.181-b13, mixed mode)

#在 mynode3 上执行如下命令,验证是否配置 JDK
[root@mynode3 ~]# java -version
java version "1.8.0_181"
Java(TM) SE Runtime Environment (build 1.8.0_181-b13)
Java HotSpot(TM) 64-Bit Server VM (build 25.181-b13, mixed mode)

#在 mynode4 上执行如下命令,验证是否配置 JDK
[root@mynode4 ~]# java -version
java version "1.8.0_181"
Java(TM) SE Runtime Environment (build 1.8.0_181-b13)
Java HotSpot(TM) 64-Bit Server VM (build 25.181-b13, mixed mode)
```

```

#在 mynode5 上执行如下命令,验证是否配置 JDK
[root@mynode5 ~]# java -version
java version "1.8.0_181"
Java(TM) SE Runtime Environment (build 1.8.0_181-b13)
Java HotSpot(TM) 64-Bit Server VM (build 25.181-b13, mixed mode)
```

通过以上操作，确定每台节点都安装了 JDK8。

### 5.3.1　节点免密

在 Kafka 中，存储消息的基本单位是分区 Partition，为了保证数据的可靠性，分区 Partition 在多个节点之间存在备份。在消息接收的过程中，消息是直接存入某个节点的 Partition 上，副本之间的数据会自动在节点之间备份，这样就需保证每台节点之间有免密登录，避免由于节点之间不能通信而导致复制数据失败。

同样，在 ZooKeeper 集群中，Follower 从节点与 Leader 节点之间也有信息传递，Follower 会上报各自的状态及当 Leader 失败时投票选举新的节点等。所以，每台节点之间也需要免密。

集群中 mynode1、mynode2、mynode3、mynode4、mynode5 两两节点之间免密步骤如下。

①在每台节点上执行命令“ssh-keygen -t rsa -P ''”，生成密钥对：

```
#在 mynode1 节点上执行如下命令
[root@mynode1 ~]# ssh-keygen -t rsa -P ''
Generating public/private rsa key pair.
Enter file in which to save the key (/root/.ssh/id_rsa):
Your identification has been saved in /root/.ssh/id_rsa.
Your public key has been saved in /root/.ssh/id_rsa.pub.
The key fingerprint is:
ce:57:f4:69:2d:58:12:e7:7a:a7:b8:27:67:e6:d1:6e root@mynode1
The key's randomart image is:
+--[ RSA 2048]----+
|..              |
|+               |
|o o             |
|. * o           |
|S + * o         |
|o . + =         |
```

```
|o . . o .          |
| . . =oE           |
| .B...             |
+-----------------+
#在 mynode2 节点上执行如下命令
[root@mynode2 ~]# ssh-keygen -t rsa -P ''
Generating public/private rsa key pair.
Enter file in which to save the key (/root/.ssh/id_rsa):
Your identification has been saved in /root/.ssh/id_rsa.
Your public key has been saved in /root/.ssh/id_rsa.pub.
The key fingerprint is:
69:77:e8:58:83:5a:02:8d:f6:31:61:83:99:62:80:c2 root@mynode2
The key's randomart image is:
+--[ RSA 2048]----+
|+. + +           |
|oEo + + o        |
|. . .+ +         |
| . o oo .        |
| o S = .         |
|= = o            |
| . . .           |
|                 |
|                 |
+-----------------+
#在 mynode3 节点上执行如下命令:
[root@mynode3 ~]# ssh-keygen -t rsa -P ''
Generating public/private rsa key pair.
Enter file in which to save the key (/root/.ssh/id_rsa):
Your identification has been saved in /root/.ssh/id_rsa.
Your public key has been saved in /root/.ssh/id_rsa.pub.
The key fingerprint is:
b4:00:ae:b0:fd:db:68:ea:1d:5c:7e:a9:40:02:3b:ac root@mynode3
The key's randomart image is:
+--[ RSA 2048]----+
| .               |
| . .             |
```

```
|o . . .          |
|.* . o .         |
|+.+ . . S        |
|..=o .           |
|E=.o             |
| ..* o           |
| .oo+ o          |
+-----------------+

#在 mynode4 节点上执行结果
[root@mynode4 ~]# ssh-keygen -t rsa -P ''
Generating public/private rsa key pair.
Enter file in which to save the key (/root/.ssh/id_rsa):
Your identification has been saved in /root/.ssh/id_rsa.
Your public key has been saved in /root/.ssh/id_rsa.pub.
The key fingerprint is:
ee:92:be:3e:26:3f:17:8b:32:fc:70:0b:91:b3:95:87 root@mynode4
The key's randomart image is:
+--[ RSA 2048]----+
|                 |
|                 |
|                 |
| .o              |
| + E S           |
|=o.              |
| .+ .o.o         |
| = +Boo          |
| XB*.            |
+-----------------+

#在 mynode5 节点上执行结果
[root@mynode5 ~]# ssh-keygen -t rsa -P ''
Generating public/private rsa key pair.
Enter file in which to save the key (/root/.ssh/id_rsa):
Your identification has been saved in /root/.ssh/id_rsa.
Your public key has been saved in /root/.ssh/id_rsa.pub.
The key fingerprint is:
```

```
b2:80:b9:c6:a6:09:91:c7:1a:a2:7f:bc:a7:56:31:8f root@mynode5
The key's randomart image is:
+--[ RSA 2048]----+
|                 |
|                 |
|                 |
|o oo             |
|= = . . = S      |
|o *  . .Eo.      |
|+  = . ..        |
|. *  + .         |
|o  .oo +         |
+-----------------+
```

②在每个节点上执行如下命令，将公钥文件写入授权文件中，并赋予权限：

```
#在 mynode1 节点上执行如下命令
[root@mynode1 ~]# cat ~/.ssh/id_rsa.pub >> ~/.ssh/authorized_keys
[root@mynode1 ~]# chmod 600 ~/.ssh/authorized_keys

#在 mynode2 节点上执行如下命令
[root@mynode2 ~]# cat ~/.ssh/id_rsa.pub >> ~/.ssh/authorized_keys
[root@mynode2 ~]# chmod 600 ~/.ssh/authorized_keys

#在 mynode3 节点上执行如下命令
[root@mynode3 ~]# cat ~/.ssh/id_rsa.pub >> ~/.ssh/authorized_keys
[root@mynode3 ~]# chmod 600 ~/.ssh/authorized_keys

#在 mynode4 节点上执行如下命令
[root@mynode4 ~]# cat ~/.ssh/id_rsa.pub >> ~/.ssh/authorized_keys
[root@mynode4 ~]# chmod 600 ~/.ssh/authorized_keys

#在 mynode5 节点上执行如下命令
[root@mynode5 ~]# cat ~/.ssh/id_rsa.pub >> ~/.ssh/authorized_keys
[root@mynode5 ~]# chmod 600 ~/.ssh/authorized_keys
```

③配置 mynode1 节点无密码登录其他节点，执行命令如下：

```
#在 mynode1 节点上执行如下命令
[root@mynode1 ~]# scp ~/.ssh/id_rsa.pub root@mynode2:~
root@mynode2's password: #输入对应节点密码
id_rsa.pub 100% 394 0.4KB/s 00:00
[root@mynode1 ~]# scp ~/.ssh/id_rsa.pub root@mynode3:~
root@mynode3's password: #输入对应节点密码
id_rsa.pub 100% 394 0.4KB/s 00:00
[root@mynode1 ~]# scp ~/.ssh/id_rsa.pub root@mynode4:~
root@mynode4's password: #输入对应节点密码
id_rsa.pub 100% 394 0.4KB/s 00:00
[root@mynode1 ~]# scp ~/.ssh/id_rsa.pub root@mynode5:~
root@mynode5's password: #输入对应节点密码
id_rsa.pub 100% 394 0.4KB/s 00:00

#在 mynode2 节点上执行如下命令
[root@mynode2 ~]# cat ~/id_rsa.pub >> ~/.ssh/authorized_keys #追加公钥到认证文件
[root@mynode2 ~]# chmod 600 ~/.ssh/authorized_keys #修改认证文件权限
[root@mynode2 ~]# rm -rf ~/id_rsa.pub #删除当前节点上的公钥

#在 mynode3 节点上执行如下命令
[root@mynode3 ~]# cat ~/id_rsa.pub >> ~/.ssh/authorized_keys
[root@mynode3 ~]# chmod 600 ~/.ssh/authorized_keys
[root@mynode3 ~]# rm -rf ~/id_rsa.pub

#在 mynode4 节点上执行如下命令
[root@mynode4 ~]# cat ~/id_rsa.pub >> ~/.ssh/authorized_keys
[root@mynode4 ~]# chmod 600 ~/.ssh/authorized_keys
[root@mynode4 ~]# rm -rf ~/id_rsa.pub

#在 mynode5 节点上执行如下命令
[root@mynode5 ~]# cat ~/id_rsa.pub >> ~/.ssh/authorized_keys
[root@mynode5 ~]# chmod 600 ~/.ssh/authorized_keys
[root@mynode5 ~]# rm -rf ~/id_rsa.pub
```

④配置 mynode2 节点无密码登录其他节点，执行命令如下：

```
#在 mynode2 节点上执行如下命令
[root@mynode2 ~]# scp ~/.ssh/id_rsa.pub root@mynode1:~
root@mynode1's password: #输入对应节点密码
id_rsa.pub 100% 394 0.4KB/s 00:00
[root@mynode2 ~]# scp ~/.ssh/id_rsa.pub root@mynode3:~
root@mynode3's password: #输入对应节点密码
id_rsa.pub 100% 394 0.4KB/s 00:00
[root@mynode2 ~]# scp ~/.ssh/id_rsa.pub root@mynode4:~
root@mynode4's password: #输入对应节点密码
id_rsa.pub 100% 394 0.4KB/s 00:00
[root@mynode2 ~]# scp ~/.ssh/id_rsa.pub root@mynode5:~
root@mynode5's password: #输入对应节点密码
id_rsa.pub 100% 394 0.4KB/s 00:00

#在 mynode1 节点上执行如下命令
[root@mynode1 ~]# cat ~/id_rsa.pub >> ~/.ssh/authorized_keys #追加公钥到认证文件
[root@mynode1 ~]# chmod 600 ~/.ssh/authorized_keys #修改认证文件权限
[root@mynode1 ~]# rm -rf ~/id_rsa.pub #删除当前节点上的公钥

#在 mynode3 节点上执行如下命令
[root@mynode3 ~]# cat ~/id_rsa.pub >> ~/.ssh/authorized_keys
[root@mynode3 ~]# chmod 600 ~/.ssh/authorized_keys
[root@mynode3 ~]# rm -rf ~/id_rsa.pub

#在 mynode4 节点上执行如下命令
[root@mynode4 ~]# cat ~/id_rsa.pub >> ~/.ssh/authorized_keys
[root@mynode4 ~]# chmod 600 ~/.ssh/authorized_keys
[root@mynode4 ~]# rm -rf ~/id_rsa.pub

#在 mynode5 节点上执行如下命令
[root@mynode5 ~]# cat ~/id_rsa.pub >> ~/.ssh/authorized_keys
[root@mynode5 ~]# chmod 600 ~/.ssh/authorized_keys
[root@mynode5 ~]# rm -rf ~/id_rsa.pub
```

⑤配置 mynode3 节点无密码登录其他节点，执行命令如下：

```
#在 mynode3 节点上执行如下命令
[root@mynode3 ~]# scp ~/.ssh/id_rsa.pub root@mynode1:~
```

```
root@mynode1's password: #输入对应节点密码
id_rsa.pub 100%  394 0.4KB/s 00:00
[root@mynode3 ~]# scp ~/.ssh/id_rsa.pub root@mynode2: ~
root@mynode2's password: #输入对应节点密码
id_rsa.pub 100%  394 0.4KB/s 00:00
[root@mynode3 ~]# scp ~/.ssh/id_rsa.pub root@mynode4: ~
root@mynode4's password: #输入对应节点密码
id_rsa.pub 100%  394 0.4KB/s 00:00
[root@mynode3 ~]# scp ~/.ssh/id_rsa.pub root@mynode5: ~
root@mynode5's password: #输入对应节点密码
id_rsa.pub 100%  394 0.4KB/s 00:00

#在 mynode1 节点上执行如下命令
[root@mynode1 ~]# cat ~/id_rsa.pub >> ~/.ssh/authorized_keys #追加公钥到认证文件
[root@mynode1 ~]# chmod 600 ~/.ssh/authorized_keys #修改认证文件权限
[root@mynode1 ~]# rm -rf ~/id_rsa.pub #删除当前节点上的公钥

#在 mynode2 节点上执行如下命令
[root@mynode2 ~]# cat ~/id_rsa.pub >> ~/.ssh/authorized_keys
[root@mynode2 ~]# chmod 600 ~/.ssh/authorized_keys
[root@mynode2 ~]# rm -rf ~/id_rsa.pub

#在 mynode4 节点上执行如下命令
[root@mynode4 ~]# cat ~/id_rsa.pub >> ~/.ssh/authorized_keys
[root@mynode4 ~]# chmod 600 ~/.ssh/authorized_keys
[root@mynode4 ~]# rm -rf ~/id_rsa.pub

#在 mynode5 节点上执行如下命令
[root@mynode5 ~]# cat ~/id_rsa.pub >> ~/.ssh/authorized_keys
[root@mynode5 ~]# chmod 600 ~/.ssh/authorized_keys
[root@mynode5 ~]# rm -rf ~/id_rsa.pub
```

⑥配置 mynode4 节点无密码登录其他节点，执行命令如下：

```
#在 mynode4 节点上执行如下命令
[root@mynode4 ~]# scp ~/.ssh/id_rsa.pub root@mynode1: ~
root@mynode1's password: #输入对应节点密码
```

```
id_rsa.pub 100%  394 0.4KB/s 00:00
[root@mynode4 ~]# scp ~/.ssh/id_rsa.pub root@mynode2:~
root@mynode2's password: #输入对应节点密码
id_rsa.pub 100%  394 0.4KB/s 00:00
[root@mynode4 ~]# scp ~/.ssh/id_rsa.pub root@mynode3:~
root@mynode3's password: #输入对应节点密码
id_rsa.pub 100%  394 0.4KB/s 00:00
[root@mynode4 ~]# scp ~/.ssh/id_rsa.pub root@mynode5:~
root@mynode5's password: #输入对应节点密码
id_rsa.pub 100%  394 0.4KB/s 00:00

#在 mynode1 节点上执行如下命令
[root@mynode1 ~]# cat ~/id_rsa.pub >> ~/.ssh/authorized_keys #追加公钥到认证文件
[root@mynode1 ~]# chmod 600 ~/.ssh/authorized_keys #修改认证文件权限
[root@mynode1 ~]# rm -rf ~/id_rsa.pub #删除当前节点上的公钥

#在 mynode2 节点上执行如下命令
[root@mynode2 ~]# cat ~/id_rsa.pub >> ~/.ssh/authorized_keys
[root@mynode2 ~]# chmod 600 ~/.ssh/authorized_keys
[root@mynode2 ~]# rm -rf ~/id_rsa.pub

#在 mynode3 节点上执行如下命令
[root@mynode3 ~]# cat ~/id_rsa.pub >> ~/.ssh/authorized_keys
[root@mynode3 ~]# chmod 600 ~/.ssh/authorized_keys
[root@mynode3 ~]# rm -rf ~/id_rsa.pub

#在 mynode5 节点上执行如下命令
[root@mynode5 ~]# cat ~/id_rsa.pub >> ~/.ssh/authorized_keys
[root@mynode5 ~]# chmod 600 ~/.ssh/authorized_keys
[root@mynode5 ~]# rm -rf ~/id_rsa.pub
```

⑦配置 mynode5 节点无密码登录其他节点，执行命令如下：

```
#在 mynode5 节点上执行如下命令
[root@mynode5 ~]# scp ~/.ssh/id_rsa.pub root@mynode1:~
root@mynode1's password: #输入对应节点密码
id_rsa.pub 100%  394 0.4KB/s 00:00
```

```
[root@mynode5 ~]# scp ~/.ssh/id_rsa.pub root@mynode2:~
root@mynode2's password: #输入对应节点密码
id_rsa.pub 100% 394 0.4KB/s 00:00
[root@mynode5 ~]# scp ~/.ssh/id_rsa.pub root@mynode3:~
root@mynode3's password: #输入对应节点密码
id_rsa.pub 100% 394 0.4KB/s 00:00
[root@mynode5 ~]# scp ~/.ssh/id_rsa.pub root@mynode4:~
root@mynode4's password: #输入对应节点密码
id_rsa.pub 100% 394 0.4KB/s 00:00

#在 mynode1 节点上执行如下命令
[root@mynode1 ~]# cat ~/id_rsa.pub >> ~/.ssh/authorized_keys #追加公钥到认证文件
[root@mynode1 ~]# chmod 600 ~/.ssh/authorized_keys #修改认证文件权限
[root@mynode1 ~]# rm -rf ~/id_rsa.pub #删除当前节点上的公钥

#在 mynode2 节点上执行如下命令
[root@mynode2 ~]# cat ~/id_rsa.pub >> ~/.ssh/authorized_keys
[root@mynode2 ~]# chmod 600 ~/.ssh/authorized_keys
[root@mynode2 ~]# rm -rf ~/id_rsa.pub

#在 mynode3 节点上执行如下命令
[root@mynode3 ~]# cat ~/id_rsa.pub >> ~/.ssh/authorized_keys
[root@mynode3 ~]# chmod 600 ~/.ssh/authorized_keys
[root@mynode3 ~]# rm -rf ~/id_rsa.pub

#在 mynode4 节点上执行如下命令
[root@mynode4 ~]# cat ~/id_rsa.pub >> ~/.ssh/authorized_keys
[root@mynode4 ~]# chmod 600 ~/.ssh/authorized_keys
[root@mynode4 ~]# rm -rf ~/id_rsa.pub
```

⑧验证节点之间免密，命令如下：

```
#在 mynode1 节点上执行如下命令
[root@mynode1 ~]# ssh mynode2
Last login: Mon Apr 22 22:36:37 2019 from 192.168.179.1
[root@mynode2 ~]# exit
logout
```

```
Connection to mynode2 closed.
[root@mynode1 ~]# ssh mynode3
Last login: Mon Apr 22 22:36:37 2019 from 192.168.179.1
[root@mynode3 ~]# exit
logout
Connection to mynode3 closed.
[root@mynode1 ~]# ssh mynode4
Last login: Mon Apr 22 22:36:38 2019 from 192.168.179.1
e[root@mynode4 ~]# exit
logout
Connection to mynode4 closed.
[root@mynode1 ~]# ssh mynode5
Last login: Mon Apr 22 23:17:25 2019 from 192.168.179.1
[root@mynode5 ~]# exit
logout
Connection to mynode5 closed.

#在 mynode2 节点上执行如下命令
[root@mynode2 ~]# ssh mynode1
Last login: Mon Apr 22 22:36:37 2019 from 192.168.179.1
[root@mynode1 ~]# exit
logout
Connection to mynode1 closed.
[root@mynode2 ~]# ssh mynode3
Last login: Mon Apr 22 23:41:01 2019 from 192.168.179.13
[root@mynode3 ~]# exit
logout
Connection to mynode3 closed.
[root@mynode2 ~]# ssh mynode4
Last login: Mon Apr 22 23:41:12 2019 from 192.168.179.13
[root@mynode4 ~]# exit
logout
Connection to mynode4 closed.
[root@mynode2 ~]# ssh mynode5
eLast login: Mon Apr 22 23:41:18 2019 from 192.168.179.13
[root@mynode5 ~]# exit
logout
Connection to mynode5 closed.
```

```

#在 mynode3 节点上执行如下命令
[root@mynode3 ~]# ssh mynode1
Last login: Mon Apr 22 23:41:23 2019 from 192.168.179.14
[root@mynode1 ~]# exit
logout
Connection to mynode1 closed.
[root@mynode3 ~]# ssh mynode2
Last login: Mon Apr 22 23:40:54 2019 from 192.168.179.13
[root@mynode2 ~]# exit
logout
Connection to mynode2 closed.
[root@mynode3 ~]# ssh mynode4
eLast login: Mon Apr 22 23:41:39 2019 from 192.168.179.14
[root@mynode4 ~]# exit
logout
Connection to mynode4 closed.
[root@mynode3 ~]# ssh mynode5
Last login: Mon Apr 22 23:41:44 2019 from 192.168.179.14
[root@mynode5 ~]# exit
logout
Connection to mynode5 closed.

#在 mynode4 节点上执行如下命令
[root@mynode4 ~]# ssh mynode1
Last login: Mon Apr 22 23:41:51 2019 from 192.168.179.15
[root@mynode1 ~]# exit
logout
Connection to mynode1 closed.
[root@mynode4 ~]# ssh mynode2
Last login: Mon Apr 22 23:41:56 2019 from 192.168.179.15
[root@mynode2 ~]# exit
logout
Connection to mynode2 closed.
[root@mynode4 ~]# ssh mynode3
Last login: Mon Apr 22 23:41:30 2019 from 192.168.179.14
[root@mynode3 ~]# exit
logout
```

```
Connection to mynode3 closed.
[root@mynode4 ~]# ssh mynode5
Last login: Mon Apr 22 23:42:05 2019 from 192.168.179.15
[root@mynode5 ~]# exit
logout
Connection to mynode5 closed.

#在 mynode5 节点上执行如下命令
[root@mynode5 ~]# ssh mynode1
Last login: Mon Apr 22 23:42:15 2019 from 192.168.179.16
[root@mynode1 ~]# exit
logout
Connection to mynode1 closed.
[root@mynode5 ~]# ssh mynode2
Last login: Mon Apr 22 23:42:20 2019 from 192.168.179.16
[root@mynode2 ~]# exit
logout
Connection to mynode2 closed.
[root@mynode5 ~]# ssh mynode3
Last login: Mon Apr 22 23:42:25 2019 from 192.168.179.16
[root@mynode3 ~]# exit
logout
Connection to mynode3 closed.
[root@mynode5 ~]# ssh mynode4
Last login: Mon Apr 22 23:42:01 2019 from 192.168.179.15
[root@mynode4 ~]# exit
logout
Connection to mynode4 closed.
```

至此，mynode1、mynode2、mynode3、mynode4、mynode5 五个节点之间已经两两免密登录，下面可以搭建 ZooKeeper 集群。

### 5.3.2 安装 ZooKeeper 协调系统

安装 Kafka 需要 ZooKeeper 分布式协调系统，如果安装过 ZooKeeper，可以忽略以下 ZooKeeper 的安装。

安装 ZooKeeper 节点的信息和 5.2.1 节划分的节点一致，安装 ZooKeeper 分布式协调系统的步骤如下。

①登录 ZooKeeper 官网，下载 ZooKeeper，这里选择的版本是 zookeeper－3.4.5，下载地址如下：

http://archive.apache.org/dist/zookeeper/zookeeper-3.4.5/

②将下载好的 zookeeper－3.4.5 上传至 mynode3 节点/software 路径下，进行解压：

```
[root@mynode3 software]# tar -zxvf ./zookeeper-3.4.5.tar.gz
```

③在 mynode3 节点上配置 ZooKeeper 的环境变量，操作如下：

```
#编辑/etc/profile 文件,在文件最后追加如下内容,并保存
[root@mynode3 software]# vim /etc/profile
export ZOOKEEPER_HOME=/software/zookeeper-3.4.5
export PATH=$PATH:$ZOOKEEPER_HOME/bin

#使 profile 文件配置生效
[root@mynode3 software]# source /etc/profile
```

④在 ZooKeeper 路径/software/zookeeper－3.4.5/conf 下配置 zoo.cfg 文件，操作如下：

```
#进入对应路径,修改 zoo.cfg 名称
[root@mynode3 conf]# cd /software/zookeeper-3.4.5/conf/
[root@mynode3 conf]# mv zoo_sample.cfgzoo.cfg

#编辑 zoo.cfg 文件
[root@mynode3 conf]# vim zoo.cfg

#编辑内容如下:
tickTime=2000
dataDir=/opt/data/zookeeper
clientPort=2181
initLimit=5
syncLimit=2
server.1=mynode3:2888:3888
server.2=mynode4:2888:3888
server.3=mynode5:2888:3888
```

⑤将以上配置好的信息发送到 mynode4、mynode5 节点，并按照上面的方式配置 ZooKeeper 的环境变量：

```
[root@mynode3 zookeeper-3.4.5]# cd /software/
[root@mynode3 software]#scp -r ./zookeeper-3.4.5 root@mynode4:'pwd'
[root@mynode3 software]#scp -r ./zookeeper-3.4.5 root@mynode5:'pwd'
```

⑥在 mynode3、mynode4、mynode5 三个节点上分别创建路径/opt/data/zookeeper 存放 ZooKeeper 数据：

```
#在 mynode3 上创建路径
[root@mynode3 software]# mkdir -p /opt/data/zookeeper

#在 mynode4 上创建路径
[root@mynode4 software]# mkdir -p /opt/data/zookeeper

#在 mynode5 上创建路径
[root@mynode5 software]# mkdir -p /opt/data/zookeeper
```

⑦在 mynode3、mynode4、mynode5 三个节点创建节点 ID，在每台节点配置的 DataDir 路径/opt/data/zookeeper 中添加 myid 文件，操作如下：

```
#在 mynode3 上路径/opt/data/zookeeper 中创建 myid 文件,并写入1
[root@mynode3 software]# cd /opt/data/zookeeper/
[root@mynode3 zookeeper]# vim myid
#写入1 保存即可

#在 mynode4 上路径/opt/data/zookeeper 中创建 myid 文件,并写入2
[root@mynode4 software]# cd /opt/data/zookeeper/
[root@mynode4 zookeeper]# vim myid #写入2 保存即可

#在 mynode5 上路径/opt/data/zookeeper 中创建 myid 文件,并写入3
[root@mynode5 software]# cd /opt/data/zookeeper/
[root@mynode5 zookeeper]# vim myid #写入3 保存即可
```

⑧在每台节点中启动 ZooKeeper 集群并验证每台节点 ZooKeeper 的角色，验证 ZooKeeper 是否安装成功：

```
#在 mynode3 启动 ZooKeeper
[root@mynode3 zookeeper]# zkServer.sh start
JMX enabled by default
Using config: /software/zookeeper-3.4.5/bin/../conf/zoo.cfg
Starting zookeeper ... STARTED

#在 mynode4 启动 ZooKeeper
[root@mynode4 zookeeper]# zkServer.sh start
JMX enabled by default
Using config: /software/zookeeper-3.4.5/bin/../conf/zoo.cfg
```

```
Starting zookeeper ... STARTED

#在 mynode5 启动 ZooKeeper
[root@mynode5 zookeeper]# zkServer.sh start
JMX enabled by default
Using config: /software/zookeeper-3.4.5/bin/../conf/zoo.cfg
Starting zookeeper ... STARTED

#在 mynode3 验证 ZooKeeper 的状态
[root@mynode3 zookeeper]# zkServer.sh status
JMX enabled by default
Using config: /software/zookeeper-3.4.5/bin/../conf/zoo.cfg
Mode: follower

#在 mynode4 验证 ZooKeeper 的状态
[root@mynode4 zookeeper]# zkServer.sh status
JMX enabled by default
Using config: /software/zookeeper-3.4.5/bin/../conf/zoo.cfg
Mode: leader

#在 mynode5 验证 ZooKeeper 的状态
[root@mynode5 zookeeper]# zkServer.sh status
JMX enabled by default
Using config: /software/zookeeper-3.4.5/bin/../conf/zoo.cfg
Mode: follower
```

通过以上步骤，ZooKeeper 成功安装。

## 任务 5.4　Kafka 集群搭建

安装 Kafka 集群之前，先登录 Kafka 官网下载 Kafka，下载网址为 http://kafka.apache.org/downloads，这里选择 Kafka_2.11-0.11.0.3 版本。下载好的 Kafka 文件名称为 kafka_2.11-0.11.0.3.tgz。将下载好的 Kafka 安装包上传到 mynode1 节点的/software/路径下进行解压，命令如下：

```
#进入/software 路径下
[root@mynode1 software]# cd /software/
#解压 Kafka 安装包
```

```
[root@mynode1 software]# tar -zxvf ./kafka_2.11-0.11.0.3.tgz
#查看解压后的安装包
[root@mynode1 software]#ll
drwxr-xr-x. 6 root root 4096 6 月 23 2018 kafka_2.11-0.11.0.3
```

### 5.4.1 节点配置

解压好 Kafka 安装包之后，首先在 mynode1 上做 Broker 的配置，然后将配置好的安装包分发到其他 Broker 节点。

要配置 mynode1 上的 Kafka 安装包，需要配置 mynode1 节点路径/software/kafka_2.11-0.11.0.3/config 下的 server.properties 文件。server.properties 详细内容如下：

```
# Licensed to the Apache Software Foundation (ASF) under one or more
# contributor license agreements. See the NOTICE file distributed with
# this work for additional information regarding copyright ownership.
# The ASF licenses this file to You under the Apache License, Version 2.0
# (the "License"); you may not use this file except in compliance with
# the License. You may obtain a copy of the License at
#
# http://www.apache.org/licenses/LICENSE-2.0
#
# Unless required by applicable law or agreed to in writing, software
# distributed under the License is distributed on an "AS IS" BASIS,
# WITHOUT WARRANTIES OR CONDITIONS OF ANY KIND, either express or implied.
# See the License for the specific language governing permissions and
# limitations under the License.

# see kafka.server.KafkaConfig for additional details and defaults

############################# Server Basics #############################
```

```

# The id of the broker. This must be set to a unique integer for each broker.
broker.id = 0 #Kafka brokerId 唯一的非负整数 id 进行标识

# Switch to enable topic deletion or not, default value is false
#delete.topic.enable = true

############################# Socket Server Settings #############################

# The address the socket server listens on. It will get the value returned from
# java.net.InetAddress.getCanonicalHostName() if not configured.
# FORMAT:
# listeners = listener_name:/host_name:port
# EXAMPLE:
# listeners = PLAINTEXT:/your.host.name:9092
#listeners = PLAINTEXT:/:9092

# Hostname and port the broker will advertise to producers and consumers. If not set,
# it uses the value for "listeners" if configured. Otherwise, it will use the value
# returned from java.net.InetAddress.getCanonicalHostName().
#advertised.listeners = PLAINTEXT:/your.host.name:9092
# Maps listener names to security protocols, the default is for them to be the same. See the config documentation for more details
#listener.security.protocol.map = PLAINTEXT:PLAINTEXT,SSL:SSL,SASL_PLAINTEXT:SASL_PLAINTEXT,SASL_SSL:SASL_SSL

# The number of threads that the server uses for receiving requests from the network and sending responses to the network
num.network.threads = 3
#broker 处理消息的最大线程数,一般情况下不需要修改

```

```
# The number of threads that the server uses for processing requests, which may include disk I/O
num.io.threads=8 # broker处理磁盘I/O的线程数,数值应该大于节点的硬盘数
# The send buffer (SO_SNDBUF) used by the socket server
socket.send.buffer.bytes=102400# socket的发送缓冲区大小
# The receive buffer (SO_RCVBUF) used by the socket server
socket.receive.buffer.bytes=102400 # socket的接收缓冲区大小
# The maximum size of a request that the socket server will accept (protection against OOM)
socket.request.max.bytes=104857600 # socket请求的最大数值,防止serverOOM
############################# Log Basics #############################
# A comma seperated list of directories under which to store log files
log.dirs=/kafka-logs #kafka数据的存放地址,多个地址的话用逗号分割
# The default number of log partitions per topic. More partitions allow greater
# parallelism for consumption, but this will also result in more files across
# the brokers.
num.partitions=1# 每个topic的分区个数,若是在topic创建时没有指定,会被topic创建时的指定参数覆盖
# The number of threads per data directory to be used for log recovery at startup and flushing at shutdown.
# This value is recommended to be increased for installations with data dirs located in RAID array.
num.recovery.threads.per.data.dir=1 #每个数据目录用于日志恢复的线程数目
```

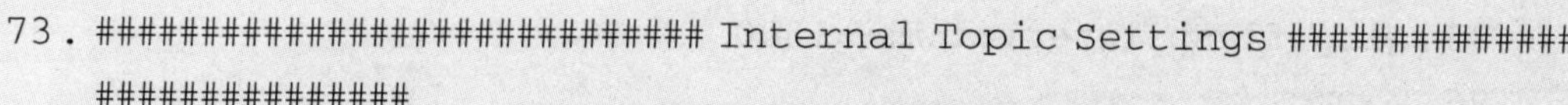



```
############################# Internal Topic Settings #############################
# The replication factor for the group metadata internal topics "__consumer_offsets" and "__transaction_state"
# For anything other than development testing, a value greater than 1 is recommended for to ensure availability such as 3.
offsets.topic.replication.factor=1# topic 的 offset 的备份份数
transaction.state.log.replication.factor=1 #事务 topic 的复制因子
transaction.state.log.min.isr=1 #覆盖事务 topic 的 min.insync.replicas 配置

############################# Log Flush Policy #############################

# Messages are immediately written to the filesystem but by default we only fsync() to sync
# the OS cache lazily. The following configurations control the flush of data to disk.
# There are a few important trade-offs here:
#1. Durability: Unflushed data may be lost if you are not using replication.
# 2. Latency: Very large flush intervals may lead to latency spikes when the flush does occur as there will be a lot of data to flush.
#3. Throughput: The flush is generally the most expensive operation, and a small flush interval may lead to exceeessive seeks.
# The settings below allow one to configure the flush policy to flush data after a period of time or
# every N messages (or both). This can be done globally and overridden on a per-topic basis.

# The number of messages to accept before forcing a flush of data to disk
#log.flush.interval.messages=10000

# The maximum amount of time a message can sit in a log before we force a flush
```

```
#log.flush.interval.ms =1000

############################### Log Retention Policy #############################

# The following configurations control the disposal of log segments. The policy can
# be set to delete segments after a period of time, or after a given size has accumulated.
# A segment will be deleted whenever *either* of these criteria are met. Deletion always happens
# from the end of the log.

# The minimum age of a log file to be eligible for deletion due to age
log.retention.hours =168#每个日志文件删除之前保存的时间

# A size-based retention policy for logs. Segments are pruned from the log as long as the remaining
# segments don't drop below log.retention.bytes. Functions independently of log.retention.hours.
#log.retention.bytes =1073741824

# The maximum size of a log segment file. When this size is reached a new log segment will be created.
#topic 分区的日志存放在某个目录下诸多文件中,这些文件将 partition 的日志切分成一段一段的,这就是段文件(segment file);一个 topic 的一个分区对应的所有 segment 文件称为 log。这个设置控制着一个 segment 文件的最大值,如果超过了这个值,就会生成一个新的 segment 文件。此配置可以被覆盖
log.segment.bytes =1073741824
# The interval at which log segments are checked to see if they can be deleted according
# to the retention policies
#检查日志段文件的间隔时间,以确定文件属性是否到达删除要求
log.retention.check.interval.ms =300000

############################### ZooKeeper ###############################
```

```
# ZooKeeper connection string (see zookeeper docs for details).
# This is a comma separated host:port pairs, each corresponding to a zk
# server. e.g. "127.0.0.1:3000,127.0.0.1:3001,127.0.0.1:3002".
# You can also append an optional chroot string to the urls to specify the
# root directory for all kafkaznodes.
zookeeper.connect = mynode3:2181,mynode4:2181,mynode5:2181 #设置 ZooKeeper 连接集群
# Timeout in ms for connecting to zookeeper
zookeeper.connection.timeout.ms = 6000 #设置 ZooKeeper 连接超时时间
################################# Group Coordinator Settings ##############################
# The following configuration specifies the time, in milliseconds, that the GroupCoordinator will delay the initial consumer rebalance.
# The rebalance will be further delayed by the value of group.initial.rebalance.delay.ms as new members join the group, up to a maximum of max.poll.interval.ms.
# The default value for this is 3 seconds.
# We override this to 0 here as it makes for a better out-of-the-box experience for development and testing.
# However, in production environments the default value of 3 seconds is more suitable as this will help to avoid unnecessary, and potentially expensive, rebalances during application startu
p.
#在执行第一次再平衡之前,group 协调员将等待更多消费者加入 group 的时间
group.initial.rebalance.delay.ms = 0
```

上面配置中每个配置参数都有详细解释，其中最核心的配置如下：

①broker. id：每个 Broker 在 Kafka 集群中都有唯一的标识，这个配置要求是正数，且每个节点之间要唯一。

②log. dir：Kafka 接收数据后，数据存储的位置。将上述配置中 Kafka 接收数据的存储位置设置在/kafka - logs 目录下，后面会检查日志数据是否写入成功。

③zookeeper. connect：设置连接 ZooKeeper 的集群，后期保存元数据装填。当搭建好的 KafKa 集群启动后，检查 ZooKeeper 中是否有元数据信息。

至此，在 mynode1 节点上的 Kafka 配置文件 server. properties 已经配置完成。

### 5.4.2 集群脚本配置

搭建好 Kafka 集群后，启动 Kafka 每个节点上的 Broker 时，每个节点 Broker 需要单独启动，并且启动后会占用前台终端，导致不能做其他的操作，这时就需要用后台去开启一个守护进程来运行 Broker 进程。在 mynode1 节点上配置启动 Kafka 集群 Broker 的脚本，后台启动每台 Broker 节点进程。创建 startkafka. sh 文件：

```
1. #在 mynode1 执行如下命令,在 Kafka 安装包下创建 startkafka.sh 文件
2. [root@mynode1 kafka_2.11 -0.11.0.3]# cd /software/kafka_2.11 -
   0.11.0.3
3. [root@mynode1 kafka_2.11 -0.11.0.3]# vim startkafka.sh
4.
5. #在文件 startkafka.sh 中配置如下内容:
6. nohup bin/kafka - server - start.sh config/server.properties >
   kafka.log 2 >&1 &
```

修改 startkafka. sh 文件权限，命令如下：

```
1. [root@mynode1 kafka_2.11 -0.11.0.3]#chmod +x ./startkafka.sh
```

此时，Kafka 集群 Broker 的启动脚本已经配置完成，下面将 mynode1 节点上的 Kafka 安装包分发到 Kafka 集群其他的 Broker 节点上。

### 5.4.3 分发安装包

在 mynode1 节点上配置好 Kafka 安装包之后，将配置好的 Kafka 安装包分发到 mynode2、mynode3 节点上。由于每台 Broker 节点的 broker. id 不同，将 Kafka 安装包分发到 mynode2、mynode3 节点之后，还需在每台节点修改对应的 server. properties 文件，配置每台节点中的 broker. id。

首先，将 mynode1 节点上的 Kafka 安装包分发到其他 Broker 节点：

```
1. #进入 mynode1 节点的 software 目录下:
2. [root@mynode1 software]# cd /software/
3.
4. #将 Kafka 安装包发送到 mynode2 节点上:
5. [root@mynode1 software]# scp -r ./kafka_2.11 -0.11.0.3 mynode2:/
   software/
```

```

#将 Kafka 安装包发送到 mynode3 节点上:
[root@mynode1 software]# scp -r ./kafka_2.11-0.11.0.3 mynode3:/software/
```

然后，修改 mynode2、mynode3 节点上每台 Broker 节点的 id：

```
#进入 mynode2 节点的 kafka 目录下:
[root@mynode2 config]# cd /software/kafka_2.11-0.11.0.3/config/

#修改 server.properties 文件
[root@mynode2 config]# vim server.properties

#在 server.properties 中找到 broker.id 并配置 broker.id=1
broker.id=1

#进入 mynode3 节点的 kafka 目录下:
[root@mynode3 config]# cd /software/kafka_2.11-0.11.0.3/config/

#修改 server.properties 文件
[root@mynode3 config]# vim server.properties

#在 server.properties 中找到 broker.id 并配置 broker.id=2
broker.id=2
```

通过配置，mynode1、mynode2、mynode3 节点上都有对应的 Kafka 安装包，一切准备就绪，就可以启动 Kafka 集群了。

### 5.4.4 集群启动

要启动 Kafka 集群，需要首先启动 ZooKeeper 集群，然后在每台 Broker 节点中启动 Broker，验证 Kafka 集群是否启动成功。成功启动 Kafka 集群需要以下两个步骤。

**1. 启动 ZooKeeper 集群**

在 mynode3、mynode4、mynode5 中启动 ZooKeeper。

```
#在 mynode3 节点执行如下命令
[root@mynode3 ~]# cd /software/zookeeper-3.4.6/bin
[root@mynode3 bin]# ./zkServer.sh start
JMX enabled by default
Using config: /software/zookeeper-3.4.6/bin/../conf/zoo.cfg
Starting zookeeper ... STARTED
```

```

#在 mynode4 节点执行如下命令
[root@mynode4 ~]# cd /software/zookeeper-3.4.6/bin
[root@mynode4 bin]# ./zkServer.sh start
JMX enabled by default
Using config: /software/zookeeper-3.4.6/bin/../conf/zoo.cfg
Starting zookeeper ... STARTED

#在 mynode5 节点执行如下命令
[root@mynode5 ~]# cd /software/zookeeper-3.4.6/bin
[root@mynode5 bin]# ./zkServer.sh start
JMX enabled by default
Using config: /software/zookeeper-3.4.6/bin/../conf/zoo.cfg
Starting zookeeper ... STARTED
```

验证 ZooKeeper 集群是否启动成功。

```
#在 mynode3 节点执行如下命令
[root@mynode3 bin]# cd /software/zookeeper-3.4.6/bin/
[root@mynode3 bin]# ./zkServer.sh status
JMX enabled by default
Using config: /software/zookeeper-3.4.6/bin/../conf/zoo.cfg
Mode: follower

#在 mynode4 节点执行如下命令
[root@mynode4 bin]# cd /software/zookeeper-3.4.6/bin/
[root@mynode4 bin]# ./zkServer.sh status
JMX enabled by default
Using config: /software/zookeeper-3.4.6/bin/../conf/zoo.cfg
Mode: leader

#在 mynode5 节点执行如下命令
[root@mynode5 bin]# cd /software/zookeeper-3.4.6/bin/
[root@mynode5 bin]# ./zkServer.sh status
JMX enabled by default
Using config: /software/zookeeper-3.4.6/bin/../conf/zoo.cfg
Mode: follower
```

通过以上命令验证，ZooKeeper 集群启动成功。

**2. 在每台 Broker 节点上启动 Kafka**

分别在 mynode1、mynode2、mynode3 节点上启动 Kafka。

```
#在 mynode1 节点执行如下命令
[root@mynode1 kafka_2.11-0.11.0.3]# cd /software/kafka_2.11-0.11.0.3
[root@mynode1 kafka_2.11-0.11.0.3]# ./startkafka.sh

#在 mynode2 节点执行如下命令
[root@mynode2 kafka_2.11-0.11.0.3]# cd /software/kafka_2.11-0.11.0.3
[root@mynode2 kafka_2.11-0.11.0.3]# ./startkafka.sh

#在 mynode3 节点执行如下命令
[root@mynode3 kafka_2.11-0.11.0.3]# cd /software/kafka_2.11-0.11.0.3
[root@mynode3 kafka_2.11-0.11.0.3]# ./startkafka.sh
```

验证 Kafka 进程。

```
#在 mynode1 节点执行如下命令
[root@mynode1 kafka_2.11-0.11.0.3]# jps
1347 Kafka #kafka 进程
1647 Jps

#在 mynode2 节点执行如下命令
[root@mynode2 kafka_2.11-0.11.0.3]# jps
1524 Kafka #kafka 进程
1822 Jps

#在 mynode3 节点执行如下命令
[root@mynode3 kafka_2.11-0.11.0.3]# jps
1380 QuorumPeerMain
1463 Kafka #kafka 进程
1719 Jps
```

通过以上命令验证，Kafka 集群启动成功。

# 项目六
# Kafka 集群测试

【项目描述】

上一项目学习了 Kafka 集群的安装，本项目将对 Kafka 原理及 Kafka 集群命令的使用进行学习。

【项目分析】

在项目建设过程中，可以使用搭建好的 Kafka 集群为 ×× 系统做数据缓存，使用 Kafka 的操作命令为 ×× 系统的稳定提供更好的数据保障。因此掌握 Kafka 集群的架构原理、Kafka 应用场景、创建 Kafka Topic 的命令、操作 Kafka Topic 的命令、Kafka 保证数据可靠的机制等是下面要学习和了解的知识。

## 任务 6.1　分布式消息系统

### 6.1.1　Kafka 介绍

Kafka 是分布式发布 - 订阅消息系统。它最初由 LinkedIn 公司开发，之后成为 Apache 项目的一部分。Kafka 是一个分布式的、可划分的、冗余备份的持久性的日志服务。它主要用于处理活跃的流式数据。在大数据系统中，常常会碰到一个问题，即整个大数据由各个子系统组成，数据需要在各个子系统中高性能、低延迟地不停地流转。传统的企业消息系统并不是很适合大规模的数据处理。为了同时处理在线应用（消息）和离线应用（数据文件、日志），出现了 Kafka。Kafka 可以起到以下两个作用：

①降低系统组网复杂度。

②降低编程复杂度。各个子系统不再是相互协商接口，各个子系统类似插口插在插座上，Kafka 承担高速数据总线的作用。

与其他的消息系统相比，Kafka 分布式消息系统在设计上综合考虑了以下方面。与其他消息系统相比，Kafka 更适合在大数据场景中使用。

**1. 吞吐量**

在分布式消息处理中，高吞吐量是必备的条件。高吞吐量是 Kafka 需要实现的核心目标之一，为了满足高吞吐量的需求，Kafka 将数据磁盘持久化，消息不在内存中缓存，直接写入磁盘，充分利用磁盘的顺序读写性能。Kafka 内部还使用零拷贝技术来减少磁盘的 I/O 操作。在向 Kafka 生产消息时，可以选择批量生产消息，同时，还支持数据的压缩，此外，还支持并行生产。上述这些特点决定了 Kafka 消息系统的高吞吐量。

**2. 负载均衡**

在 Kafka 分布式消息系统架构中，消息是写往消息队列中的，也就是写往 Topic，每个 Topic 由很多个分区组成，这些分区的出现满足并行地向 Kafka 中生产消息的需求。用户可以根据需求来指定消息写往哪个分区，这样可以使消息均匀分布在 Kafka 中。

同时，为保证数据的高可靠性，每个 Topic 的分区还有备份，将这些备份分散到不同的节点上，每个分区都由一台 Kafka 节点管理，由 ZooKeeper 保证故障恢复。假设某台 Kafka 节点挂掉，可以由 ZooKeeper 协调将当前节点管理的 Partition 转交其他 Kafka 节点管理，这样可以实现高可用性。同时，ZooKeeper 还可以协调 Kafka 系统中动态地加入 Kafka 节点及消费者，自动实现均衡。

**3. 拉取系统**

由于 Kafka Broker 会持久化数据，Kafka 采用了零拷贝技术，Broker 没有内存压力，因此，消费者非常适合采取 pull 的方式消费数据，这样就简化了 Kafka 设计。消费者根据消费能力自主控制消息拉取速度，自主选择消费模式，例如批量、重复消费、从尾端开始消费等。

**4. 可扩展性**

当需要增加 Broker 结点时，新增的 Broker 会向 ZooKeeper 注册，而生产者及消费者会根据注册在 ZooKeeper 上的 Watcher 感知这些变化，并及时做出调整。

### 6.1.2　Kafka 架构

Kafka 的整体架构非常简单，是显式分布式架构。Kafka 是生产者/消费者模式，向 Kafka 中生产消息的一端叫作生产者，生产者可以有多个，这样可以由多个生产者向 Kafka 中生产消息。消息写往 Kafka 中的 Broker 节点，由于 Kafka 中采用了零拷贝技术，消息直接写到 Broker 节点的磁盘上。Broker 节点也可以有多个，这样为 Kafka 集群的高可用及分布式提供了保证。消费 Kafka 消息的一端叫作消费者，消费者也可以有多个，这样可以保证消息的并行消费，这也是 Kafka 分布式消息系统更适用于大数据场景的原因之一。

数据从生产者发送到 Broker，Broker 承担着中间缓存和分发的作用，类似于缓存，即活跃的数据和离线处理系统之间的缓存。ZooKeeper 在 Kafka 集群中起着协调作用，保存 Kafka 集群的元数据信息，可以为 Kafka 的故障恢复提供保障。Kafka 集群的结构图如图 6－1 所示。

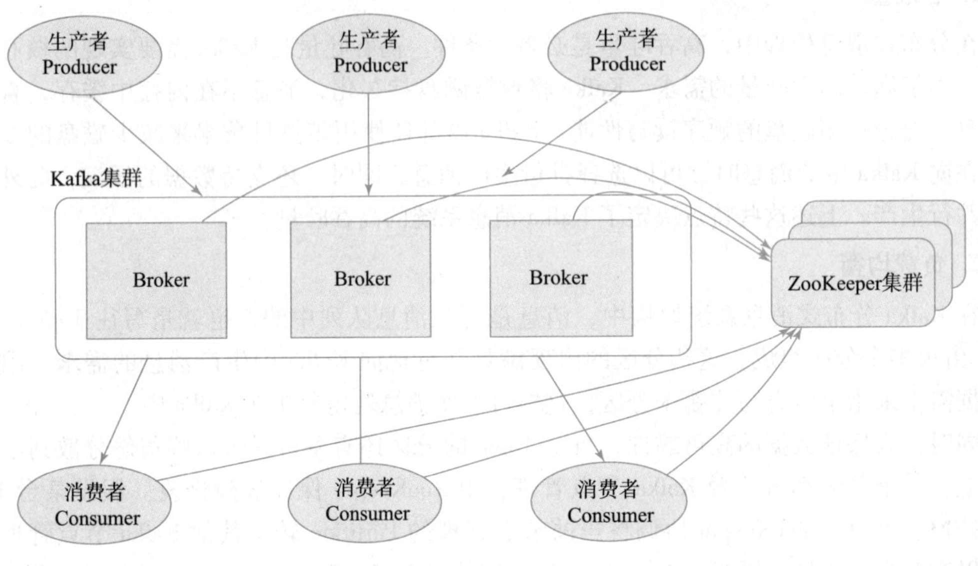


图 6－1　Kafka 架构图

Kafka 中有如下几个概念。

**1. Producer**

Producer 是 Kafka 分布式消息系统的消息生产者，自己决定将消息写入 Kafka 中 Topic 的哪个分区。生产者生产消息时，消息可以是 K,V 格式。默认生产者生产消息时，消息的 Key 为空，消息是轮询写往某个 Topic 中的。如果生产者的消息指定 K，那么默认消息写往 Partition 的方式就是基于 K 的 Hash 值与 Topic 中 Partition 的个数取模，每条消息按照这种规则写入某个 Partition 中。

**2. Borker**

Broker 是组成 Kafka 集群的节点，一个 Kafka 集群中可以有多个 Broker。生产者生产的消息是直接写到 Broker 节点的磁盘上的，每个 Broker 可以管理多个 Topic 的 Partition，负责 Partition 消息的读写和存储。Broker 之间是没有主从关系的，当一个 Broker 挂掉后，当前 Broker 节点上管理的 Partition 由 ZooKeeper 协调管理并寻找新的 Broker 节点来管理。

**3. Topic**

生产者将消息写往 Topic。Topic 就是消息队列，一类消息的总称。Kafka 将消息写往不同的 Topic 的主要原因是将不同类别的消息归类到不同的 Topic 中，这样也方便后期消费者消费。Topic 是由 Partition 组成的，可以在创建 Topic 时指定当前 Topic 由几个 Partition 组成。Topic 由多个 Partition 组成主要是为了解决消息的并行消费问题，每个 Partition 可以同时被一个消费者消费，那么多个 Partition 就可以并行地为多个消费者同时消费。

**4. Partition**

Partition 是组成 Kafka Topic 的基本单元，写往 Kafka 中的消息是直接写到 Partition 上的。

Partition 对应 Broker 节点上的一个目录，内部消息是强有序的。消息写到 Partition 上是直接追加写入的，所以在 Partition 内部消息，是严格的 FIFO，这样在同一个 Partition 中也保证的消息的一致性。

每个 Partition 还有副本，这些副本分布在不同的 Broker 节点上。每个 Partition 都由一个 Broker 节点来管理，这个 Borker 节点就叫作这个 Partition 的 Leader。当前这个Partition 的副本会自动同步 Leader 节点管理的 Partition 的数据，保证各个 Partition 副本数据的一致性。

**5. Consumer**

Consumer 是消息的消费者，每个 Consumer 都有各自的消费者组的概念。不同的消费者组消费同一个 Topic 中的数据时，它们之间互不影响。同一个消费者组内有不同的消费者，这些消费者消费同一个 Topic 消息时，这个 Topic 中的相同的消息只能被消费一次。

Topic 中的某个 Partition 只能同时被同一个消费者组内的消费者连接消费，这样如果某个 Topic 的分区数为 10，那么这个 Topic 可以并行地被一个消费者组内的 10 个消费者同时连接消费，这样就大大增加消费的效率。

**6. ZooKeeper**

ZooKeeper 在 Kafka 集群中主要有两个作用：第一，可以存储 Kafka 集群的一些元数据信息，这些数据信息包括 Kafka 集群的节点，Kafka 集群的 Topic 信息、Broker 信息等。第二，在 Kafka 的老版本中还负责存储消费者消息的 offset，例如在 Kafka 0.8.2 版本之前，所有消费者消费的消息的 offset 都是存储在 ZooKeeper 中的。在 Kafka 0.8.2 之后的版本，消息的 offset 默认由 Kafka 管理。

### 6.1.3 Kafka 的特点

Kafka 集群的特点如下。

①Kafka 消息系统同时为发布和订阅提供高吞吐量。据了解，Kafka 每秒可以生产约 25 万消息（50 MB），每秒处理 55 万消息（110 MB）。

②可进行持久化操作。将消息持久化到磁盘，因此可用于批量消费，例如 ETL，以及实时应用程序。通过将数据持久化到硬盘及 Replication，防止数据丢失。

③分布式系统，易于向外扩展。所有的 Producer、Broker 和 Consumer 都会有多个，均为分布式的。无须停机即可扩展机器。

④消息被处理的状态是在 Consumer 端维护，而不是在 Server 端。当失败时，能自动平衡。

⑤支持 online 和 offline 的场景。

### 6.1.4 Kafka 应用场景

Kafka 分布式消息系统的应用场景如下。

**1. 消息队列**

比起大多数的消息系统来说，Kafka 有更好的吞吐量、内置的分区、冗余及容错性，这让 Kafka 成为一个很好的大规模消息处理应用的解决方案。消息系统一般吞吐量相对较低，但是需要更小的端到端延时，并常常依赖于 Kafka 提供的强大的持久性保障。在这个领域，Kafka 足以媲美传统消息系统，如 ActiveMR 或 RabbitMQ。

**2. 行为跟踪**

Kafka 的另一个应用场景是跟踪用户浏览页面、搜索及其他行为，以发布 - 订阅的模式实时记录到对应的 Topic 里。这些结果被订阅者拿到后，就可以做进一步的实时处理或实时监控，或放到 adoop、离线数据仓库里处理。

**3. 元信息监控**

作为操作记录的监控模块来使用，即汇集记录一些操作信息，这里可以理解为运维性质的数据监控。

**4. 日志收集**

日志收集方面，其实开源产品有很多，包括 Scribe、Apache Flume。很多人使用 Kafka 代替日志聚合（log aggregation）。日志聚合一般来说是从服务器上收集日志文件，然后放到一个集中的位置（文件服务器或 HDFS）进行处理。然而 Kafka 忽略掉文件的细节，将其更清晰地抽象成一个个日志或事件的消息流。这就让 Kafka 处理过程延迟更低，更容易支持多数据源和分布式数据处理。比起以日志为中心的系统比如 Scribe 或者 Flume 来说，Kafka 提供同样高效的性能和由于复制导致的更高的耐用性保证，以及更低的端到端延迟。

**5. 流处理**

这个场景可能比较多，也很好理解。保存收集流数据，以供之后对接的 Storm 或其他流式计算框架进行处理。很多用户会将那些从原始 Topic 来的数据进行阶段性处理、汇总、扩充，或者以其他方式转换到新的 Topic 下再继续后面的处理。例如，一个文章推荐的处理流程，可能是先从 RSS 数据源中抓取文章的内容，然后将其丢入一个叫作“文章”的 Topic 中，后续操作可能是对这个内容进行清理，比如恢复正常数据或者删除重复数据，最后再将内容匹配的结果返还给用户。这就在一个独立的 Topic 之外，产生了一系列的实时数据处理的流程。Strom 和 Samza 是非常著名的实现这种类型数据转换的框架。

**6. 事件源**

事件源是一种应用程序设计的方式。该方式的状态转移被记录为按时间顺序排序的记录序列。Kafka 可以存储大量的日志数据，这使得它成为一个对这种方式的应用来说绝佳的后台。

**7. 持久性日志**

Kafka 可以为一种外部的持久性日志的分布式系统提供服务。这种日志可以在节点间备份数据，并为故障节点数据回复提供一种重新同步的机制。Kafka 中日志压缩功能为这种用法提供了条件。在这种用法中，Kafka 类似于 Apache BookKeeper 项目。

## 任务6.2 测试Kafka

通过学习，知道了Kafka分布式消息系统的架构、特点及应用场景，从本任务开始，将学习如何使用命令操作Kafka系统、如何创建消息队列、如何查询消息队列、如何生成消息及如何消费消息等。

### 6.2.1 查看集群中的Topic

搭建好Kafka集群后，在集群中默认是没有Topic的，需要创建Topic。创建Topic后，查询后才能看到创建的Topic。查询Topic时，首先需要启动Kafka集群，而启动Kafka集群需要先启动ZooKeeper集群。

①启动ZooKeeper集群，命令如下：

```
#在 mynode3 节点上启动 ZooKeeper
[root@mynode3 ~]# zkServer.sh start
JMX enabled by default
Using config: /software/zookeeper-3.4.6/bin/../conf/zoo.cfg
Starting zookeeper ... STARTED

#在 mynode4 节点上启动 ZooKeeper
[root@mynode4 ~]# zkServer.sh start
JMX enabled by default
Using config: /software/zookeeper-3.4.6/bin/../conf/zoo.cfg
Starting zookeeper ... STARTED

#在 mynode5 节点上启动 ZooKeeper
[root@mynode5 ~]# zkServer.sh start
JMX enabled by default
Using config: /software/zookeeper-3.4.6/bin/../conf/zoo.cfg
Starting zookeeper ... STARTED
```

②启动Kafka集群。要启动Kafka集群，需要在每台Broker节点上单独启动Kafka。在mynode1节点上执行命令，启动Kafka，命令如下：

```
[root@mynode1 ~]# cd /software/kafka_2.11-0.11.0.3/bin
[root@mynode1 bin]# ./kafka-server-start.sh /software/kafka_
  2.11-0.11.0.3/config/server.properties
[2019-05-07 07:25:17,102] INFO KafkaConfig values:
advertised.host.name=null
```

```
5. advertised.listeners=null
6. advertised.port=null
7. alter.config.policy.class.name=null
8. authorizer.class.name=
9. auto.create.topics.enable=true
10. auto.leader.rebalance.enable=true
11. background.threads=10
12. broker.id=0
13. broker.id.generation.enable=true
14. broker.rack=null
15. compression.type=producer
16. connections.max.idle.ms=600000
17. controlled.shutdown.enable=true
18. controlled.shutdown.max.retries=3
19. controlled.shutdown.retry.backoff.ms=5000
20. controller.socket.timeout.ms=30000
21. create.topic.policy.class.name=null
22. default.replication.factor=1
23. delete.records.purgatory.purge.interval.requests=1
24. delete.topic.enable=false
25. fetch.purgatory.purge.interval.requests=1000
26. group.initial.rebalance.delay.ms=0
27. group.max.session.timeout.ms=300000
28. group.min.session.timeout.ms=6000
29. host.name=
30. inter.broker.listener.name=null
31. inter.broker.protocol.version=0.11.0-IV2
32. leader.imbalance.check.interval.seconds=300
33. leader.imbalance.per.broker.percentage=10
34. listener.security.protocol.map=SSL:SSL,SASL_PLAINTEXT:SASL_
    PLAINTEXT,TRACE:TRACE,SASL_SSL:SASL_SSL,PLAINTEXT:PLAINTEXT
35. listeners=null
36. log.cleaner.backoff.ms=15000
37. log.cleaner.dedupe.buffer.size=134217728
38. log.cleaner.delete.retention.ms=86400000
39. log.cleaner.enable=true
40. log.cleaner.io.buffer.load.factor=0.9
41. log.cleaner.io.buffer.size=524288
```

```
42. log.cleaner.io.max.bytes.per.second=1.7976931348623157E308
43. log.cleaner.min.cleanable.ratio=0.5
44. log.cleaner.min.compaction.lag.ms=0
45. log.cleaner.threads=1
46. log.cleanup.policy=[delete]
47. log.dir=/tmp/kafka-logs
48. log.dirs=/kafka-logs
49. log.flush.interval.messages=9223372036854775807
50. log.flush.interval.ms=null
51. log.flush.offset.checkpoint.interval.ms=60000
52. log.flush.scheduler.interval.ms=9223372036854775807
53. log.flush.start.offset.checkpoint.interval.ms=60000
54. log.index.interval.bytes=4096
55. log.index.size.max.bytes=10485760
56. log.message.format.version=0.11.0-IV2
57. log.message.timestamp.difference.max.ms=9223372036854775807
58. log.message.timestamp.type=CreateTime
59. log.preallocate=false
60. log.retention.bytes=-1
61. log.retention.check.interval.ms=300000
62. log.retention.hours=168
63. log.retention.minutes=null
64. log.retention.ms=null
65. log.roll.hours=168
66. log.roll.jitter.hours=0
67. log.roll.jitter.ms=null
68. log.roll.ms=null
69. log.segment.bytes=1073741824
70. log.segment.delete.delay.ms=60000
71. max.connections.per.ip=2147483647
72. max.connections.per.ip.overrides=
73. message.max.bytes=1000012
74. metric.reporters=[]
75. metrics.num.samples=2
76. metrics.recording.level=INFO
77. metrics.sample.window.ms=30000
78. min.insync.replicas=1
79. num.io.threads=8
```

```
num.network.threads =3
num.partitions =1
num.recovery.threads.per.data.dir =1
num.replica.fetchers =1
offset.metadata.max.bytes =4096
offsets.commit.required.acks = -1
offsets.commit.timeout.ms =5000
offsets.load.buffer.size =5242880
offsets.retention.check.interval.ms =600000
offsets.retention.minutes =1440
offsets.topic.compression.codec =0
offsets.topic.num.partitions =50
offsets.topic.replication.factor =1
offsets.topic.segment.bytes =104857600
port =9092
principal.builder.class = class org.apache.kafka.common.security.auth.DefaultPrincipalBuilder
producer.purgatory.purge.interval.requests =1000
queued.max.requests =500
quota.consumer.default =9223372036854775807
quota.producer.default =9223372036854775807
quota.window.num =11
quota.window.size.seconds =1
replica.fetch.backoff.ms =1000
replica.fetch.max.bytes =1048576
replica.fetch.min.bytes =1
replica.fetch.response.max.bytes =10485760
replica.fetch.wait.max.ms =500
replica.high.watermark.checkpoint.interval.ms =5000
replica.lag.time.max.ms =10000
replica.socket.receive.buffer.bytes =65536
replica.socket.timeout.ms =30000
replication.quota.window.num =11
replication.quota.window.size.seconds =1
request.timeout.ms =30000
reserved.broker.max.id =1000
sasl.enabled.mechanisms = [GSSAPI]
sasl.kerberos.kinit.cmd = /usr/bin/kinit
```

```
sasl.kerberos.min.time.before.relogin = 60000
sasl.kerberos.principal.to.local.rules = [DEFAULT]
sasl.kerberos.service.name = null
sasl.kerberos.ticket.renew.jitter = 0.05
sasl.kerberos.ticket.renew.window.factor = 0.8
sasl.mechanism.inter.broker.protocol = GSSAPI
security.inter.broker.protocol = PLAINTEXT
socket.receive.buffer.bytes = 102400
socket.request.max.bytes = 104857600
socket.send.buffer.bytes = 102400
ssl.cipher.suites = null
ssl.client.auth = none
ssl.enabled.protocols = [TLSv1.2, TLSv1.1, TLSv1]
ssl.endpoint.identification.algorithm = null
ssl.key.password = null
ssl.keymanager.algorithm = SunX509
ssl.keystore.location = null
ssl.keystore.password = null
ssl.keystore.type = JKS
ssl.protocol = TLS
ssl.provider = null
ssl.secure.random.implementation = null
ssl.trustmanager.algorithm = PKIX
ssl.truststore.location = null
ssl.truststore.password = null
ssl.truststore.type = JKS
transaction.abort.timed.out.transaction.cleanup.interval.ms = 60000
transaction.max.timeout.ms = 900000
transaction.remove.expired.transaction.cleanup.interval.ms = 3600000
transaction.state.log.load.buffer.size = 5242880
transaction.state.log.min.isr = 1
transaction.state.log.num.partitions = 50
transaction.state.log.replication.factor = 1
transaction.state.log.segment.bytes = 104857600
transactional.id.expiration.ms = 604800000
unclean.leader.election.enable = false
```

```
zookeeper.connect=mynode3:2181,mynode4:2181,mynode5:2181
zookeeper.connection.timeout.ms=6000
zookeeper.session.timeout.ms=6000
zookeeper.set.acl=false
zookeeper.sync.time.ms=2000
(kafka.server.KafkaConfig)
[2019-05-07 07:25:17,216] INFO starting (kafka.server.KafkaServer)
[2019-05-07 07:25:17,237] INFO Connecting to zookeeper on mynode3:2181,mynode4:2181,mynode5:2181 (kafka.server.KafkaServer)
[2019-05-07 07:25:17,257] INFO Starting ZkClient event thread. (org.I0Itec.zkclient.ZkEventThread)
[2019-05-07 07:25:17,271] INFO Client environment:zookeeper.version=3.4.10-39d3a4f269333c922ed3db283be479f9deacaa0f, built on 03/23/2017 10:13 GMT (org.apache.zookeeper.ZooKeeper)
[2019-05-07 07:25:17,271] INFO Client environment:host.name=mynode1 (org.apache.zookeeper.ZooKeeper)
[2019-05-07 07:25:17,271] INFO Client environment:java.version=1.8.0_181 (org.apache.zookeeper.ZooKeeper)
[2019-05-07 07:25:17,271] INFO Client environment:java.vendor=Oracle Corporation (org.apache.zookeeper.ZooKeeper)
[2019-05-07 07:25:17,271] INFO Client environment:java.home=/software/jdk1.8.0_181/jre (org.apache.zookeeper.ZooKeeper)
[2019-05-07 07:25:17,271] INFO Client environment:java.class.path=:/software/kafka_2.11-0.11.0.3/bin/../libs/aopalliance-repackaged-2.5.0-b05.jar:/software/kafka_2.11-0.11.0.3/bin/../libs/a
rgparse4j-0.7.0.jar:/software/kafka_2.11-0.11.0.3/bin/../libs/commons-lang3-
3.5.jar:/software/kafka_2.11-0.11.0.3/bin/../libs/connect-api-
0.11.0.3.jar:/software/kafka_2.11-0.11.0.3/bin/../libs/connect-file-
0.11.0.3.jar:/software/kafka_2.11-0.11.0.3/bin/../libs/connect-json-
0.11.0.3.jar:/software/kafka_2.11-0.11.0.3/bin/../libs/connect-runtime-
```

```
0.11.0.3.jar:/software/kafka_2.11-0.11.0.3/bin/../libs/con-
nect-transforms-
0.11.0.3.jar:/software/kafka_2.11-0.11.0.3/bin/../libs/
guava-
20.0.jar:/software/kafka_2.11-0.11.0.3/bin/../libs/hk2-
api-2.5.0-
b05.jar:/software/kafka_2.11-0.11.0.3/bin/../libs/hk2-lo-
cator-2.5.0-
b05.jar:/software/kafka_2.11-0.11.0.3/bin/../libs/hk2-
utils-2.5.0-
b05.jar:/software/kafka_2.11-0.11.0.3/bin/../libs/jackson-
annotations-
2.8.11.jar:/software/kafka_2.11-0.11.0.3/bin/../libs/jack-
son-core-
2.8.10.jar:/software/kafka_2.11-0.11.0.3/bin/../libs/jack-
son-core-
2.8.11.jar:/software/kafka_2.11-0.11.0.3/bin/../libs/jack-
son-databind-
2.8.11.2.jar:/software/kafka_2.11-0.11.0.3/bin/../libs/
jackson-jaxrs-base-
2.8.11.jar:/software/kafka_2.11-0.11.0.3/bin/../libs/jack-
son-jaxrs-json-provider-
2.8.11.jar:/software/kafka_2.11-0.11.0.3/bin/../libs/jack-
son-module-jaxb-annotations-
2.8.11.jar:/software/kafka_2.11-0.11.0.3/bin/../libs/javas-
sist-3.21.0-
GA.jar:/software/kafka_2.11-0.11.0.3/bin/../libs/javax.an-
notation-api-
1.2.jar:/software/kafka_2.11-0.11.0.3/bin/../libs/javax.in-
ject-
1.jar:/software/kafka_2.11-0.11.0.3/bin/../libs/javax.in-
ject-2.5.0-
b05.jar:/software/kafka_2.11-0.11.0.3/bin/../libs/javax.
servlet-api-
3.1.0.jar:/software/kafka_2.11-0.11.0.3/bin/../libs/javax.
ws.rs-api-
2.0.1.jar:/software/kafka_2.11-0.11.0.3/bin/../libs/jersey-
client-
```

```
2.24.jar:/software/kafka_2.11-0.11.0.3/bin/../libs/jersey-
common-
2.24.jar:/software/kafka_2.11-0.11.0.3/bin/../libs/jersey-
container-servlet-
2.24.jar:/software/kafka_2.11-0.11.0.3/bin/../libs/jersey-
container-servlet-core-
2.24.jar:/software/kafka_2.11-0.11.0.3/bin/../libs/jersey-
guava-
2.24.jar:/software/kafka_2.11-0.11.0.3/bin/../libs/jersey-
media-jaxb-
2.24.jar:/software/kafka_2.11-0.11.0.3/bin/../libs/jersey-
server-
2.24.jar:/software/kafka_2.11-0.11.0.3/bin/../libs/jetty-
continuation-
9.2.22.v20170606.jar:/software/kafka_2.11-0.11.0.3/bin/../
libs/jetty-http-
9.2.22.v20170606.jar:/software/kafka_2.11-0.11.0.3/bin/../
libs/jetty-io-
9.2.22.v20170606.jar:/software/kafka_2.11-0.11.0.3/bin/../
libs/jetty-security-
9.2.22.v20170606.jar:/software/kafka_2.11-0.11.0.3/bin/../
libs/jetty-server-
9.2.22.v20170606.jar:/software/kafka_2.11-0.11.0.3/bin/../
libs/jetty-servlet-
9.2.22.v20170606.jar:/software/kafka_2.11-0.11.0.3/bin/../
libs/jetty-servlets-
9.2.22.v20170606.jar:/software/kafka_2.11-0.11.0.3/bin/../
libs/jetty-util-
9.2.22.v20170606.jar:/software/kafka_2.11-0.11.0.3/bin/../
libs/jopt-simple-
5.0.3.jar:/software/kafka_2.11-0.11.0.3/bin/../libs/kafka_
2.11-
0.11.0.3.jar:/software/kafka_2.11-0.11.0.3/bin/../libs/kaf-
ka_2.11-0.11.0.3-
sources.jar:/software/kafka_2.11-0.11.0.3/bin/../libs/kaf-
ka_2.11-0.11.0.3-test-
sources.jar:/software/kafka_2.11-0.11.0.3/bin/../libs/kaf-
ka-clients-
```

```
0.11.0.3.jar:/software/kafka_2.11-0.11.0.3/bin/../libs/kaf-
ka-log4j-appender-
0.11.0.3.jar:/software/kafka_2.11-0.11.0.3/bin/../libs/kaf-
ka-streams-
0.11.0.3.jar:/software/kafka_2.11-0.11.0.3/bin/../libs/kaf-
ka-streams-examples-
0.11.0.3.jar:/software/kafka_2.11-0.11.0.3/bin/../libs/kaf-
ka-tools-
0.11.0.3.jar:/software/kafka_2.11-0.11.0.3/bin/../libs/
log4j-
1.2.17.jar:/software/kafka_2.11-0.11.0.3/bin/../libs/lz4-
1.3.0.jar:/software/kafka_2.11-0.11.0.3/bin/../libs/maven-
artifact-
3.5.0.jar:/software/kafka_2.11-0.11.0.3/bin/../libs/met-
rics-core-
2.2.0.jar:/software/kafka_2.11-0.11.0.3/bin/../libs/osgi-
resource-locator-
1.0.1.jar:/software/kafka_2.11-0.11.0.3/bin/../libs/plexus-
utils-
3.0.24.jar:/software/kafka_2.11-0.11.0.3/bin/../libs/
reflections-
0.9.11.jar:/software/kafka_2.11-0.11.0.3/bin/../libs/rocks-
dbjni-
5.0.1.jar:/software/kafka_2.11-0.11.0.3/bin/../libs/scala-
library-
2.11.11.jar:/software/kafka_2.11-0.11.0.3/bin/../libs/sca-
la-parser-combinators_2.11-
1.0.4.jar:/software/kafka_2.11-0.11.0.3/bin/../libs/slf4j-
api-
1.7.25.jar:/software/kafka_2.11-0.11.0.3/bin/../libs/slf4j-
log4j12-
1.7.25.jar:/software/kafka_2.11-0.11.0.3/bin/../libs/snap-
py-java-
1.1.2.6.jar:/software/kafka_2.11-0.11.0.3/bin/../libs/vali-
dation-api-
1.1.0.Final.jar:/software/kafka_2.11-0.11.0.3/bin/../libs/
zkclient-
```

```
    0.10.jar:/software/kafka_2.11-0.11.0.3/bin/../libs/zookeeper-
    3.4.10.jar
    (org.apache.zookeeper.ZooKeeper)[2019-05-07 07:25:17,272]
    INFO Client
    environment:java.library.path=/usr/java/packages/lib/
    amd64:/usr/lib64:/lib64:/lib:/usr/lib (org.apache.zookeep-
    er.ZooKeeper)
169.[2019-05-07 07:25:17,272] INFO Client environment:java.io.
    tmpdir=/tmp (org.apache.zookeeper.ZooKeeper)
170.[2019-05-07 07:25:17,272] INFO Client environment:java.com-
    piler=<NA> (org.apache.zookeeper.ZooKeeper)
171.[2019-05-07 07:25:17,272] INFO Client environment:os.name=
    Linux (org.apache.zookeeper.ZooKeeper)
172.[2019-05-07 07:25:17,272] INFO Client environment:os.arch=
    amd64 (org.apache.zookeeper.ZooKeeper)
173.[2019-05-07 07:25:17,272] INFO Client environment:os.version
    =2.6.32-504.el6.x86_64 (org.apache.zookeeper.ZooKeeper)
174.......
175.......
176.[2019-05-07 07:25:20,290] INFO Replica loaded for partition
    flumetopic-0 with initial high watermark 0 (kafka.cluster.
    Replica)
177.[2019-05-07 07:25:20,290] INFO Replica loaded for partition
    flumetopic-0 with initial high watermark 0 (kafka.cluster.
    Replica)
178.[2019-05-07 07:25:20,290] INFO Partition [__consumer_off-
    sets,13] on broker 0: __consumer_offsets-13 starts at Leader
    Epoch 21 from offset 0. Previous Leader Epoch was: 20 (kafka.
    cluster.Partition)
179.[2019-05-07 07:25:20,366] INFO [GroupCoordinator 0]: Loading
    group metadata for console-consumer-48707 with generation 2
    (kafka.coordinator.group.GroupCoordinator)
180.[2019-05-07 07:25:20,367] INFO [Group Metadata Manager on
    Broker 0]: Finished loading offsets and group metadata from __
    consumer_offsets-37 in 80 milliseconds. (kafka.coordinator.
    group.GroupMetadataManager)
181.[2019-05-07 07:25:20,367] INFO [Group Metadata Manager on
    Broker 0]: Finished loading offsets and group metadata from __
```

```
consumer_offsets - 40 in 0 milliseconds. (kafka.coordinator.
group.Group MetadataManager)
182.[2019 - 05 - 07 07:25:20,402] INFO [Group Metadata Manager on
Broker 0]: Finished loading offsets and group metadata from __
consumer_offsets - 46 in 35 milliseconds. (kafka.coordinator.
group.GroupMetadataManager)
183.[2019 - 05 - 07 07:25:20,402] INFO [Group Metadata Manager on
Broker 0]: Finished loading offsets and group metadata from __
consumer_offsets - 49 in 0 milliseconds. (kafka.coordinator.
group.Group MetadataManager)
184.[2019 -05 -07 07:25:20,448] INFO [GroupCoordinator 0]: Loading
group metadata for mydemo3Group with generation 34 (kafka.co-
ordinator.group.GroupCoordinator)
185.[2019 - 05 - 07 07:25:20,448] INFO [Group Metadata Manager on
Broker 0]: Finished loading offsets and group metadata from __
consumer_offsets - 43 in 46 milliseconds. (kafka.coordinator.
group.Grou pMetadataManager)
186.[2019 - 05 - 07 07:25:20,448] INFO [Group Metadata Manager on
Broker 0]: Finished loading offsets and group metadata from __
consumer_offsets - 1 in 0 milliseconds. (kafka.coordinator.
group.GroupMetadataManager)
187.[2019 - 05 - 07 07:25:20,448] INFO [Group Metadata Manager on
Broker 0]: Finished loading offsets and group metadata
from __ consumer_offsets -4 in 0 milliseconds. (kafka.coordinator.
group.GroupMetadataManager)
188.[2019 - 05 - 07 07:25:20,448] INFO [Group Metadata Manager on
Broker 0]: Finished loading offsets and group metadata
from __ consumer_offsets -7 in 0 milliseconds. (kafka.coordinator.
group.GroupMetadataManager)
189.[2019 - 05 - 07 07:25:20,480] INFO [Group Metadata Manager on
Broker 0]: Finished loading offsets and group metadata from __
consumer_offsets - 10 in 32 milliseconds. (kafka.coordinator.
group.GroupMetadataManager)
190.[2019 - 05 - 07 07:25:20,481] INFO [Group Metadata Manager on
Broker 0]: Finished loading offsets and group metadata from __
consumer_offsets - 13 in 0 milliseconds. (kafka.coordinator.
group.GroupMetadataManager)
```

```
[2019-05-07 07:25:20,521] INFO [Group Metadata Manager on Broker 0]: Finished loading offsets and group metadata from __consumer_offsets-16 in 40 milliseconds. (kafka.coordinator.group.GroupMetadataManager)
[2019-05-07 07:25:20,522] INFO [Group Metadata Manager on Broker 0]: Finished loading offsets and group metadata from __consumer_offsets-19 in 0 milliseconds. (kafka.coordinator.group.GroupMetadataManager)
......
......
```

在 mynode2 上单独启动 Kafka，执行如下命令：

```
[root@mynode2 ~]# cd /software/kafka_2.11-0.11.0.3/bin/
[root@mynode2 bin]# ./kafka-server-start.sh /software/kafka_2.11-0.11.0.3/config/server.properties
[2019-05-07 07:30:56,079] INFO KafkaConfig values:
advertised.host.name = null
advertised.listeners = null
advertised.port = null
alter.config.policy.class.name = null
authorizer.class.name =
auto.create.topics.enable = true
auto.leader.rebalance.enable = true
background.threads = 10
broker.id = 1
broker.id.generation.enable = true
broker.rack = null
compression.type = producer
connections.max.idle.ms = 600000
controlled.shutdown.enable = true
controlled.shutdown.max.retries = 3
controlled.shutdown.retry.backoff.ms = 5000
controller.socket.timeout.ms = 30000
create.topic.policy.class.name = null
default.replication.factor = 1
delete.records.purgatory.purge.interval.requests = 1
delete.topic.enable = false
fetch.purgatory.purge.interval.requests = 1000
```

```
26. group.initial.rebalance.delay.ms=0
27. group.max.session.timeout.ms=300000
28. group.min.session.timeout.ms=6000
29. host.name=
30. inter.broker.listener.name=null
31. inter.broker.protocol.version=0.11.0-IV2
32. leader.imbalance.check.interval.seconds=300
33. leader.imbalance.per.broker.percentage=10
34. listener.security.protocol.map=SSL:SSL,SASL_PLAINTEXT:SASL_
    PLAINTEXT,TRACE:TRACE,SASL_SSL:SASL_SSL,PLAINTEXT:PLAINTEXT
35. listeners=null
36. log.cleaner.backoff.ms=15000
37. log.cleaner.dedupe.buffer.size=134217728
38. log.cleaner.delete.retention.ms=86400000
39. log.cleaner.enable=true
40. log.cleaner.io.buffer.load.factor=0.9
41. log.cleaner.io.buffer.size=524288
42. log.cleaner.io.max.bytes.per.second=1.7976931348623157E308
43. log.cleaner.min.cleanable.ratio=0.5
44. log.cleaner.min.compaction.lag.ms=0
45. log.cleaner.threads=1
46. log.cleanup.policy=[delete]
47. log.dir=/tmp/kafka-logs
48. log.dirs=/kafka-logs
49. log.flush.interval.messages=9223372036854775807
50. log.flush.interval.ms=null
51. log.flush.offset.checkpoint.interval.ms=60000
52. log.flush.scheduler.interval.ms=9223372036854775807
53. log.flush.start.offset.checkpoint.interval.ms=60000
54. log.index.interval.bytes=4096
55. log.index.size.max.bytes=10485760
56. log.message.format.version=0.11.0-IV2
57. log.message.timestamp.difference.max.ms=9223372036854775807
58. log.message.timestamp.type=CreateTime
59. log.preallocate=false
60. log.retention.bytes=-1
61. log.retention.check.interval.ms=300000
62. log.retention.hours=168
```

```
log.retention.minutes = null
log.retention.ms = null
log.roll.hours = 168
log.roll.jitter.hours = 0
log.roll.jitter.ms = null
log.roll.ms = null
log.segment.bytes = 1073741824
log.segment.delete.delay.ms = 60000
max.connections.per.ip = 2147483647
max.connections.per.ip.overrides =
message.max.bytes = 1000012
metric.reporters = []
metrics.num.samples = 2
metrics.recording.level = INFO
metrics.sample.window.ms = 30000
min.insync.replicas = 1
num.io.threads = 8
num.network.threads = 3
num.partitions = 1
num.recovery.threads.per.data.dir = 1
num.replica.fetchers = 1
offset.metadata.max.bytes = 4096
offsets.commit.required.acks = -1
offsets.commit.timeout.ms = 5000
offsets.load.buffer.size = 5242880
offsets.retention.check.interval.ms = 600000
offsets.retention.minutes = 1440
offsets.topic.compression.codec = 0
offsets.topic.num.partitions = 50
offsets.topic.replication.factor = 1
offsets.topic.segment.bytes = 104857600
port = 9092
principal.builder.class = class org.apache.kafka.common.security.auth.DefaultPrincipalBuilder
producer.purgatory.purge.interval.requests = 1000
queued.max.requests = 500
quota.consumer.default = 9223372036854775807
quota.producer.default = 9223372036854775807
```

```
quota.window.num=11
quota.window.size.seconds=1
replica.fetch.backoff.ms=1000
replica.fetch.max.bytes=1048576
replica.fetch.min.bytes=1
replica.fetch.response.max.bytes=10485760
replica.fetch.wait.max.ms=500
replica.high.watermark.checkpoint.interval.ms=5000
replica.lag.time.max.ms=10000
replica.socket.receive.buffer.bytes=65536
replica.socket.timeout.ms=30000
replication.quota.window.num=11
replication.quota.window.size.seconds=1
request.timeout.ms=30000
reserved.broker.max.id=1000
sasl.enabled.mechanisms=[GSSAPI]
sasl.kerberos.kinit.cmd=/usr/bin/kinit
sasl.kerberos.min.time.before.relogin=60000
sasl.kerberos.principal.to.local.rules=[DEFAULT]
sasl.kerberos.service.name=null
sasl.kerberos.ticket.renew.jitter=0.05
sasl.kerberos.ticket.renew.window.factor=0.8
sasl.mechanism.inter.broker.protocol=GSSAPI
security.inter.broker.protocol=PLAINTEXT
socket.receive.buffer.bytes=102400
socket.request.max.bytes=104857600
socket.send.buffer.bytes=102400
ssl.cipher.suites=null
ssl.client.auth=none
ssl.enabled.protocols=[TLSv1.2, TLSv1.1, TLSv1]
ssl.endpoint.identification.algorithm=null
ssl.key.password=null
ssl.keymanager.algorithm=SunX509
ssl.keystore.location=null
ssl.keystore.password=null
ssl.keystore.type=JKS
ssl.protocol=TLS
ssl.provider=null
```

```
ssl.secure.random.implementation = null
ssl.trustmanager.algorithm = PKIX
ssl.truststore.location = null
ssl.truststore.password = null
ssl.truststore.type = JKS
transaction.abort.timed.out.transaction.cleanup.interval.ms = 60000
transaction.max.timeout.ms = 900000
transaction.remove.expired.transaction.cleanup.interval.ms = 3600000
transaction.state.log.load.buffer.size = 5242880
transaction.state.log.min.isr = 1
transaction.state.log.num.partitions = 50
transaction.state.log.replication.factor = 1
transaction.state.log.segment.bytes = 104857600
transactional.id.expiration.ms = 604800000
unclean.leader.election.enable = false
zookeeper.connect = mynode3:2181,mynode4:2181,mynode5:2181
zookeeper.connection.timeout.ms = 6000
zookeeper.session.timeout.ms = 6000
zookeeper.set.acl = false
zookeeper.sync.time.ms = 2000
(kafka.server.KafkaConfig)
[2019-05-07 07:30:56,207] INFO starting (kafka.server.KafkaServer)
[2019-05-07 07:30:56,229] INFO Connecting to zookeeper on mynode3:2181,mynode4:2181,mynode5:2181 (kafka.server.KafkaServer)
[2019-05-07 07:30:56,247] INFO Starting ZkClient event thread. (org.I0Itec.zkclient.ZkEventThread)
[2019-05-07 07:30:56,260] INFO Client environment:zookeeper.version=3.4.10-39d3a4f269333c922ed3db283be479f9deacaa0f, built on 03/23/2017 10:13 GMT (org.apache.zookeeper.ZooKeeper)
[2019-05-07 07:30:56,260] INFO Client environment:host.name=mynode2 (org.apache.zookeeper.ZooKeeper)
[2019-05-07 07:30:56,261] INFO Client environment:java.version=1.8.0_181 (org.apache.zookeeper.ZooKeeper)
[2019-05-07 07:30:56,261] INFO Client environment:java.vendor=Oracle Corporation (org.apache.zookeeper.ZooKeeper)
```

```
166. [2019-05-07 07:30:56,261] INFO Client environment:java.home=/
     software/jdk1.8.0_181/jre (org.apache.zookeeper.ZooKeeper)
167. [2019-05-07 07:30:56,261] INFO Client environment:java.
     class.path=:/software/kafka_2.11-0.11.0.3/bin/../libs/aop-
     alliance-repackaged-2.5.0-b05.jar:/software/kafka_2.11-
     0.11.0.3/bin/../libs/a
168. rgparse4j-0.7.0.jar:/software/kafka_2.11-0.11.0.3/bin/../
     libs/commons-lang3-
     3.5.jar:/software/kafka_2.11-0.11.0.3/bin/../libs/connect-
     api-
     0.11.0.3.jar:/software/kafka_2.11-0.11.0.3/bin/../libs/con-
     nect-file-
     0.11.0.3.jar:/software/kafka_2.11-0.11.0.3/bin/../libs/con-
     nect-json-
     0.11.0.3.jar:/software/kafka_2.11-0.11.0.3/bin/../libs/con-
     nect-runtime-
     0.11.0.3.jar:/software/kafka_2.11-0.11.0.3/bin/../libs/con-
     nect-transforms-
     0.11.0.3.jar:/software/kafka_2.11-0.11.0.3/bin/../libs/
     guava-
     20.0.jar:/software/kafka_2.11-0.11.0.3/bin/../libs/hk2-
     api-2.5.0-
     b05.jar:/software/kafka_2.11-0.11.0.3/bin/../libs/hk2-
     locator-2.5.0-
     b05.jar:/software/kafka_2.11-0.11.0.3/bin/../libs/hk2-
     utils-2.5.0-
     b05.jar:/software/kafka_2.11-0.11.0.3/bin/../libs/jackson-
     annotations-
     2.8.11.jar:/software/kafka_2.11-0.11.0.3/bin/../libs/jack-
     son-core-
     2.8.10.jar:/software/kafka_2.11-0.11.0.3/bin/../libs/jack-
     son-core-
     2.8.11.jar:/software/kafka_2.11-0.11.0.3/bin/../libs/jack-
     son-databind-
     2.8.11.2.jar:/software/kafka_2.11-0.11.0.3/bin/../libs/
     jackson-jaxrs-base-
     2.8.11.jar:/software/kafka_2.11-0.11.0.3/bin/../libs/jack-
     son-jaxrs-json-provider-
```

```
2.8.11.jar:/software/kafka_2.11-0.11.0.3/bin/../libs/jack-
son-module-jaxb-annotations-
2.8.11.jar:/software/kafka_2.11-0.11.0.3/bin/../libs/javas-
sist-3.21.0-
GA.jar:/software/kafka_2.11-0.11.0.3/bin/../libs/javax.
annotation-api-
1.2.jar:/software/kafka_2.11-0.11.0.3/bin/../libs/
javax.inject-
1.jar:/software/kafka_2.11-0.11.0.3/bin/../libs/javax.in-
ject-2.5.0-
b05.jar:/software/kafka_2.11-0.11.0.3/bin/../libs/javax.
servlet-api-
3.1.0.jar:/software/kafka_2.11-0.11.0.3/bin/../libs/javax.
ws.rs-api-
2.0.1.jar:/software/kafka_2.11-0.11.0.3/bin/../libs/jersey-
client-
2.24.jar:/software/kafka_2.11-0.11.0.3/bin/../libs/jersey-
common-
2.24.jar:/software/kafka_2.11-0.11.0.3/bin/../libs/jersey-
container-servlet-
2.24.jar:/software/kafka_2.11-0.11.0.3/bin/../libs/jersey-
container-servlet-core-
2.24.jar:/software/kafka_2.11-0.11.0.3/bin/../libs/jersey-
guava-
2.24.jar:/software/kafka_2.11-0.11.0.3/bin/../libs/jersey-
media-jaxb-
2.24.jar:/software/kafka_2.11-0.11.0.3/bin/../libs/jersey-
server-
2.24.jar:/software/kafka_2.11-0.11.0.3/bin/../libs/jetty-
continuation-
9.2.22.v20170606.jar:/software/kafka_2.11-0.11.0.3/bin/../
libs/jetty-http-
9.2.22.v20170606.jar:/software/kafka_2.11-0.11.0.3/bin/../
libs/jetty-io-
9.2.22.v20170606.jar:/software/kafka_2.11-0.11.0.3/bin/../
libs/jetty-security-
9.2.22.v20170606.jar:/software/kafka_2.11-0.11.0.3/bin/../
libs/jetty-server-
```

```
9.2.22.v20170606.jar:/software/kafka_2.11-0.11.0.3/bin/../
libs/jetty-servlet-
9.2.22.v20170606.jar:/software/kafka_2.11-0.11.0.3/bin/../
libs/jetty-servlets-
9.2.22.v20170606.jar:/software/kafka_2.11-0.11.0.3/bin/../
libs/jetty-util-
9.2.22.v20170606.jar:/software/kafka_2.11-0.11.0.3/bin/../
libs/jopt-simple-
5.0.3.jar:/software/kafka_2.11-0.11.0.3/bin/../libs/kafka_
2.11-
0.11.0.3.jar:/software/kafka_2.11-0.11.0.3/bin/../libs/
kafka_2.11-0.11.0.3-
sources.jar:/software/kafka_2.11-0.11.0.3/bin/../libs/
kafka_2.11-0.11.0.3-test-
sources.jar:/software/kafka_2.11-0.11.0.3/bin/../libs/
kafka-clients-
0.11.0.3.jar:/software/kafka_2.11-0.11.0.3/bin/../libs/
kafka-log4j-appender-
0.11.0.3.jar:/software/kafka_2.11-0.11.0.3/bin/../libs/
kafka-streams-
0.11.0.3.jar:/software/kafka_2.11-0.11.0.3/bin/../libs/
kafka-streams-examples-
0.11.0.3.jar:/software/kafka_2.11-0.11.0.3/bin/../libs/
kafka-tools-
0.11.0.3.jar:/software/kafka_2.11-0.11.0.3/bin/../libs/
log4j-
1.2.17.jar:/software/kafka_2.11-0.11.0.3/bin/../libs/lz4-
1.3.0.jar:/software/kafka_2.11-0.11.0.3/bin/../libs/maven-
artifact-
3.5.0.jar:/software/kafka_2.11-0.11.0.3/bin/../libs/
metrics-core-
2.2.0.jar:/software/kafka_2.11-0.11.0.3/bin/../libs/osgi-
resource-locator-
1.0.1.jar:/software/kafka_2.11-0.11.0.3/bin/../libs/plexus-
utils-
3.0.24.jar:/software/kafka_2.11-0.11.0.3/bin/../libs/
reflections-
```

```
0.9.11.jar:/software/kafka_2.11-0.11.0.3/bin/../libs/rocks-
dbjni-
5.0.1.jar:/software/kafka_2.11-0.11.0.3/bin/../libs/scala-
library-
2.11.11.jar:/software/kafka_2.11-0.11.0.3/bin/../libs/
scala-parser-combinators_2.11-
1.0.4.jar:/software/kafka_2.11-0.11.0.3/bin/../libs/slf4j-
api-
1.7.25.jar:/software/kafka_2.11-0.11.0.3/bin/../libs/slf4j-
log4j12-
1.7.25.jar:/software/kafka_2.11-0.11.0.3/bin/../libs/snap-
py-java-
1.1.2.6.jar:/software/kafka_2.11-0.11.0.3/bin/../libs/vali-
dation-api-
1.1.0.Final.jar:/software/kafka_2.11-0.11.0.3/bin/../libs/
zkclient-
0.10.jar:/software/kafka_2.11-0.11.0.3/bin/../libs/zookeeper-
3.4.10.jar
(org.apache.zookeeper.ZooKeeper)[2019-05-07 07:30:56,261]
INFO Client
environment:java.library.path=/usr/java/packages/lib/
amd64:/usr/lib64:/lib64:/lib:/usr/lib (org.apache.zookeep-
er.ZooKeeper)
[2019-05-07 07:30:56,261] INFO Client environment:java.io.
tmpdir=/tmp (org.apache.zookeeper.ZooKeeper)
[2019-05-07 07:30:56,261] INFO Client environment:java.com-
piler=<NA> (org.apache.zookeeper.ZooKeeper)
[2019-05-07 07:30:56,261] INFO Client environment:os.name=
Linux (org.apache.zookeeper.ZooKeeper)
[2019-05-07 07:30:56,261] INFO Client environment:os.arch=
amd64 (org.apache.zookeeper.ZooKeeper)
[2019-05-07 07:30:56,261] INFO Client environment:os.version
=2.6.32-504.el6.x86_64 (org.apache.zookeeper.ZooKeeper)
[2019-05-07 07:30:56,261] INFO Client environment:user.name
=root (org.apache.zookeeper.ZooKeeper)
[2019-05-07 07:30:56,261] INFO Client environment:user.home
=/root (org.apache.zookeeper.ZooKeeper)
```

```
176. [2019-05-07 07:30:56,261] INFO Client environment:user.dir=/software/kafka_2.11-0.11.0.3/bin (org.apache.zookeeper.ZooKeeper)
177. ......
178. ......
179. [2019-05-07 07:30:59,012] INFO Truncating flumetopic-2 to 3 has no effect as the largest offset in the log is 2. (kafka.log.Log)
180. [2019-05-07 07:30:59,012] INFO Truncating mytest-0 to 21 has no effect as the largest offset in the log is 20. (kafka.log.Log)
181. [2019-05-07 07:30:59,012] INFO Truncating flumetopic-1 to 2 has no effect as the largest offset in the log is 1. (kafka.log.Log)
182. [2019-05-07 07:30:59,012] INFO Truncating mytest-2 to 21 has no effect as the largest offset in the log is 20. (kafka.log.Log)
183. [2019-05-07 07:30:59,012] INFO Truncating flumetopic-0 to 2 has no effect as the largest offset in the log is 1. (kafka.log.Log)
184. [2019-05-07 07:30:59,013] INFO [Group Metadata Manager on Broker 1]: Finished loading offsets and group metadata from __consumer_offsets-2 in 23 milliseconds. (kafka.coordinator.group.GroupMetadataManager)
185. [2019-05-07 07:30:59,018] INFO [Group Metadata Manager on Broker 1]: Finished loading offsets and group metadata from __consumer_offsets-5 in 0 milliseconds. (kafka.coordinator.group.GroupMetadataManager)
186. [2019-05-07 07:30:59,043] INFO [ReplicaFetcherManager on broker 1] Removed fetcher for partitions __consumer_offsets-8,mytopic-0,__consumer_offsets-35,__consumer_offsets-41,__consumer_offsets-23,__consumer_offsets-47,__consumer_offsets-38,__consumer_offsets-17,__consumer_offsets-11,__consumer_offsets-2,__consumer_offsets-14,__consumer_offsets-20,__consumer_offsets-44,__consumer_offsets-5,__consumer_offsets-26,__consumer_offsets-29,__consumer_offsets-32 (kafka.server.ReplicaFetcherManager)
```

```
187. [2019-05-07 07:30:59,043] INFO Partition [mytopic,0] on bro-
     ker 1: mytopic-0 starts at Leader Epoch 4 from offset 1. Previ-
     ous Leader Epoch was: 3 (kafka.cluster.Partition)
188. [2019-05-07 07:30:59,055] INFO Partition [__consumer_off-
     sets,29] on broker 1: __consumer_offsets-29 starts at Leader
     Epoch 24 from offset 244. Previous Leader Epoch was: 23 (kafka.
     cluster.Partition)
189. [2019-05-07 07:30:59,057] INFO Partition [__consumer_off-
     sets,26] on broker 1: __consumer_offsets-26 starts at Leader
     Epoch 24 from offset 2637. Previous Leader Epoch was: 23 (kafka.
     cluster.Partition)
190. [2019-05-07 07:30:59,057] INFO Partition [__consumer_off-
     sets,23] on broker 1: __consumer_offsets-23 starts at Leader
     Epoch 24 from offset 185. Previous Leader Epoch was: 23 (kafka.
     cluster.Partition)
191. [2019-05-07 07:30:59,058] INFO Partition [__consumer_off-
     sets,20] on broker 1: __consumer_offsets-20 starts at Leader
     Epoch 24 from offset 0. Previous Leader Epoch was: 23 (kafka.
     cluster.Partition)
192. [2019-05-07 07:30:59,058] INFO Partition [__consumer_off-
     sets,17] on broker 1: __consumer_offsets-17 starts at Leader
     Epoch 24 from offset 1822. Previous Leader Epoch was: 23 (kafka.
     cluster.Partition)
193. [2019-05-07 07:30:59,058] INFO Partition [__consumer_off-
     sets,14] on broker 1: __consumer_offsets-14 starts at Leader
     Epoch 24 from offset 0. Previous Leader Epoch was: 23 (kafka.
     cluster.Partition)
194. [2019-05-07 07:30:59,059] INFO Partition [__consumer_off-
     sets,11] on broker 1: __consumer_offsets-11 starts at Leader
     Epoch 24 from offset 0. Previous Leader Epoch was: 23 (kafka.
     cluster.Partition)
195. [2019-05-07 07:30:59,060] INFO Partition [__consumer_off-
     sets,8] on broker 1: __consumer_offsets-8 starts at Leader Ep-
     och 24 from offset 97. Previous Leader Epoch was: 23 (kafka.
     cluster.Partition)
196. [2019-05-07 07:30:59,060] INFO Partition [__consumer_off-
     sets,5] on broker 1: __consumer_offsets-5 starts at Leader
```

```
    Epoch 24 from offset 0. Previous Leader Epoch was: 23 (kafka.
    cluster.Partition)
197.[2019-05-07 07:30:59,060] INFO Partition [__consumer_off-
    sets,2] on broker 1: __consumer_offsets-2 starts at Leader Ep-
    och 24 from offset 0. Previous Leader Epoch was: 23 (kafka.clus-
    ter.Parti tion)
198.[2019-05-07 07:30:59,060] INFO Partition [__consumer_off-
    sets,47] on broker 1: __consumer_offsets-47 starts at Leader
    Epoch 24 from offset 714. Previous Leader Epoch was: 23 (kafka.
    cluster.Partition)
199.[2019-05-07 07:30:59,060] INFO Partition [__consumer_off-
    sets,38] on broker 1: __consumer_offsets-38 starts at Leader
    Epoch 24 from offset 66. Previous Leader Epoch was: 23 (kafka.
    cluster.Partition)
200.[2019-05-07 07:30:59,061] INFO Partition [__consumer_off-
    sets,35] on broker 1: __consumer_offsets-35 starts at Leader
    Epoch 24 from offset 0. Previous Leader Epoch was: 23 (kafka.
    cluster.Partition)
201.[2019-05-07 07:30:59,061] INFO Partition [__consumer_off-
    sets,44] on broker 1: __consumer_offsets-44 starts at Leader
    Epoch 24 from offset 14322. Previous Leader Epoch was: 23 (kaf-
    ka.clusterPartition)
202..[2019-05-07 07:30:59,061] INFO Partition [__consumer_off-
    sets,32] on broker 1: __consumer_offsets-32 starts at Leader
    Epoch 24 from offset 0. Previous Leader Epoch was: 23 (kafka.
    cluster.Partition)
203.[2019-05-07 07:30:59,061] INFO Partition [__consumer_off-
    sets,41] on broker 1: __consumer_offsets-41 starts at Leader
    Epoch 24 from offset 332. Previous Leader Epoch was: 23 (kafka.
    cluster.Partition)
204.[2019-05-07 07:30:59,110] INFO [Group Metadata Manager on
    Broker 1]: Finished loading offsets and group metadata from __
    consumer_offsets-8 in 92 milliseconds. (kafka.coordinator.
    group.GroupMetadataManager)
205.[2019-05-07 07:30:59,111] INFO [Group Metadata Manager on
    Broker 1]: Finished loading offsets and group metadata from __
    consumer_offsets-11 in 0 milliseconds. (kafka.coordinator.
    group.GroupMetadataManager)
```

```
[2019-05-07 07:30:59,111] INFO [Group Metadata Manager on Broker 1]: Finished loading offsets and group metadata from __consumer_offsets-14 in 0 milliseconds. (kafka.coordinator.group.GroupMetadataManager)
[2019-05-07 07:30:59,177] INFO [Group Metadata Manager on Broker 1]: Finished loading offsets and group metadata from __consumer_offsets-17 in 65 milliseconds. (kafka.coordinator.group.GroupMetadataManager)
[2019-05-07 07:30:59,178] INFO [Group Metadata Manager on Broker 1]: Finished loading offsets and group metadata from __consumer_offsets-20 in 0 milliseconds. (kafka.coordinator.group.GroupMetadataManager)
[2019-05-07 07:30:59,211] INFO [Group Metadata Manager on Broker 1]: Finished loading offsets and group metadata from __consumer_offsets-23 in 33 milliseconds. (kafka.coordinator.group.GroupMetadataManager)
[2019-05-07 07:30:59,244] INFO [Group Metadata Manager on Broker 1]: Finished loading offsets and group metadata from __consumer_offsets-26 in 33 milliseconds. (kafka.coordinator.group.GroupMetadataManager)
[2019-05-07 07:30:59,278] INFO [Group Metadata Manager on Broker 1]: Finished loading offsets and group metadata from __consumer_offsets-29 in 34 milliseconds. (kafka.coordinator.group.GroupMetadataManager)
[2019-05-07 07:30:59,278] INFO [Group Metadata Manager on Broker 1]: Finished loading offsets and group metadata from __consumer_offsets-32 in 0 milliseconds. (kafka.coordinator.group.GroupMetadataManager)
[2019-05-07 07:30:59,278] INFO [Group Metadata Manager on Broker 1]: Finished loading offsets and group metadata from __consumer_offsets-35 in 0 milliseconds. (kafka.coordinator.group.GroupMetadataManager)
[2019-05-07 07:30:59,310] INFO [Group Metadata Manager on Broker 1]: Finished loading offsets and group metadata from __consumer_offsets-38 in 32 milliseconds. (kafka.coordinator.group.GroupMetadataManager)
[2019-05-07 07:30:59,344] INFO [Group Metadata Manager on Broker 1]: Finished loading offsets and group metadata from __
```

```
consumer_offsets-41 in 34 milliseconds.(kafka.coordinator.group.GroupMetadataManager)
[2019-05-07 07:30:59,414] INFO [Group Metadata Manager on Broker 1]: Finished loading offsets and group metadata from __consumer_offsets-44 in 70 milliseconds.(kafka.coordinator.group.GroupMetadataManager)
[2019-05-07 07:30:59,448] INFO [Group Metadata Manager on Broker 1]: Finished loading offsets and group metadata from __consumer_offsets-47 in 34 milliseconds.(kafka.coordinator.group.GroupMetadataManager)
......
......
```

在 mynode3 上单独启动 Kafka，执行如下命令：

```
[root@mynode3 ~]# cd /software/kafka_2.11-0.11.0.3/bin/
[root@mynode3 bin]# ./kafka-server-start.sh /software/kafka_2.11-0.11.0.3/config/server.properties
[2019-05-07 07:30:56,079] INFO KafkaConfig values:
advertised.host.name = null
advertised.listeners = null
advertised.port = null
alter.config.policy.class.name = null
authorizer.class.name =
auto.create.topics.enable = true
auto.leader.rebalance.enable = true
background.threads = 10
broker.id = 1
broker.id.generation.enable = true
broker.rack = null
compression.type = producer
connections.max.idle.ms = 600000
controlled.shutdown.enable = true
controlled.shutdown.max.retries = 3
controlled.shutdown.retry.backoff.ms = 5000
controller.socket.timeout.ms = 30000
create.topic.policy.class.name = null
default.replication.factor = 1
delete.records.purgatory.purge.interval.requests = 1
```

```
delete.topic.enable = false
fetch.purgatory.purge.interval.requests = 1000
group.initial.rebalance.delay.ms = 0
group.max.session.timeout.ms = 300000
group.min.session.timeout.ms = 6000
host.name =
inter.broker.listener.name = null
inter.broker.protocol.version = 0.11.0-IV2
leader.imbalance.check.interval.seconds = 300
leader.imbalance.per.broker.percentage = 10
listener.security.protocol.map = SSL:SSL,SASL_PLAINTEXT:SASL_PLAINTEXT,TRACE:TRACE,SASL_SSL:SASL_SSL,PLAINTEXT:PLAINTEXT
listeners = null
log.cleaner.backoff.ms = 15000
log.cleaner.dedupe.buffer.size = 134217728
log.cleaner.delete.retention.ms = 86400000
log.cleaner.enable = true
log.cleaner.io.buffer.load.factor = 0.9
log.cleaner.io.buffer.size = 524288
log.cleaner.io.max.bytes.per.second = 1.7976931348623157E308
log.cleaner.min.cleanable.ratio = 0.5
log.cleaner.min.compaction.lag.ms = 0
log.cleaner.threads = 1
log.cleanup.policy = [delete]
log.dir = /tmp/kafka-logs
log.dirs = /kafka-logs
log.flush.interval.messages = 9223372036854775807
log.flush.interval.ms = null
log.flush.offset.checkpoint.interval.ms = 60000
log.flush.scheduler.interval.ms = 9223372036854775807
log.flush.start.offset.checkpoint.interval.ms = 60000
log.index.interval.bytes = 4096
log.index.size.max.bytes = 10485760
log.message.format.version = 0.11.0-IV2
log.message.timestamp.difference.max.ms = 9223372036854775807
log.message.timestamp.type = CreateTime
log.preallocate = false
log.retention.bytes = -1
```

```
61. log.retention.check.interval.ms = 300000
62. log.retention.hours = 168
63. log.retention.minutes = null
64. log.retention.ms = null
65. log.roll.hours = 168
66. log.roll.jitter.hours = 0
67. log.roll.jitter.ms = null
68. log.roll.ms = null
69. log.segment.bytes = 1073741824
70. log.segment.delete.delay.ms = 60000
71. max.connections.per.ip = 2147483647
72. max.connections.per.ip.overrides =
73. message.max.bytes = 1000012
74. metric.reporters = []
75. metrics.num.samples = 2
76. metrics.recording.level = INFO
77. metrics.sample.window.ms = 30000
78. min.insync.replicas = 1
79. num.io.threads = 8
80. num.network.threads = 3
81. num.partitions = 1
82. num.recovery.threads.per.data.dir = 1
83. num.replica.fetchers = 1
84. offset.metadata.max.bytes = 4096
85. offsets.commit.required.acks = -1
86. offsets.commit.timeout.ms = 5000
87. offsets.load.buffer.size = 5242880
88. offsets.retention.check.interval.ms = 600000
89. offsets.retention.minutes = 1440
90. offsets.topic.compression.codec = 0
91. offsets.topic.num.partitions = 50
92. offsets.topic.replication.factor = 1
93. offsets.topic.segment.bytes = 104857600
94. port = 9092
95. principal.builder.class = class org.apache.kafka.common.secur-
    ity.auth.DefaultPrincipalBuilder
96. producer.purgatory.purge.interval.requests = 1000
97. queued.max.requests = 500
```

```
quota.consumer.default=9223372036854775807
quota.producer.default=9223372036854775807
quota.window.num=11
quota.window.size.seconds=1
replica.fetch.backoff.ms=1000
replica.fetch.max.bytes=1048576
replica.fetch.min.bytes=1
replica.fetch.response.max.bytes=10485760
replica.fetch.wait.max.ms=500
replica.high.watermark.checkpoint.interval.ms=5000
replica.lag.time.max.ms=10000
replica.socket.receive.buffer.bytes=65536
replica.socket.timeout.ms=30000
replication.quota.window.num=11
replication.quota.window.size.seconds=1
request.timeout.ms=30000
reserved.broker.max.id=1000
sasl.enabled.mechanisms=[GSSAPI]
sasl.kerberos.kinit.cmd=/usr/bin/kinit
sasl.kerberos.min.time.before.relogin=60000
sasl.kerberos.principal.to.local.rules=[DEFAULT]
sasl.kerberos.service.name=null
sasl.kerberos.ticket.renew.jitter=0.05
sasl.kerberos.ticket.renew.window.factor=0.8
sasl.mechanism.inter.broker.protocol=GSSAPI
security.inter.broker.protocol=PLAINTEXT
socket.receive.buffer.bytes=102400
socket.request.max.bytes=104857600
socket.send.buffer.bytes=102400
ssl.cipher.suites=null
ssl.client.auth=none
ssl.enabled.protocols=[TLSv1.2, TLSv1.1, TLSv1]
ssl.endpoint.identification.algorithm=null
ssl.key.password=null
ssl.keymanager.algorithm=SunX509
ssl.keystore.location=null
ssl.keystore.password=null
ssl.keystore.type=JKS
```

```
ssl.protocol = TLS
ssl.provider = null
ssl.secure.random.implementation = null
ssl.trustmanager.algorithm = PKIX
ssl.truststore.location = null
ssl.truststore.password = null
ssl.truststore.type = JKS
transaction.abort.timed.out.transaction.cleanup.interval.ms = 60000
transaction.max.timeout.ms = 900000
transaction.remove.expired.transaction.cleanup.interval.ms = 3600000
transaction.state.log.load.buffer.size = 5242880
transaction.state.log.min.isr = 1
transaction.state.log.num.partitions = 50
transaction.state.log.replication.factor = 1
transaction.state.log.segment.bytes = 104857600
transactional.id.expiration.ms = 604800000
unclean.leader.election.enable = false
zookeeper.connect = mynode3:2181,mynode4:2181,mynode5:2181
zookeeper.connection.timeout.ms = 6000
zookeeper.session.timeout.ms = 6000
zookeeper.set.acl = false
zookeeper.sync.time.ms = 2000
(kafka.server.KafkaConfig)
[2019-05-07 07:30:56,207] INFO starting (kafka.server.KafkaServer)
[2019-05-07 07:30:56,229] INFO Connecting to zookeeper on mynode3:2181,mynode4:2181,mynode5:2181 (kafka.server.KafkaServer)
[2019-05-07 07:30:56,247] INFO Starting ZkClient event thread. (org.I0Itec.zkclient.ZkEventThread)
[2019-05-07 07:30:56,260] INFO Client environment:zookeeper.version = 3.4.10-39d3a4f269333c922ed3db283be479f9deacaa0f, built on 03/23/2017 10:13 GMT (org.apache.zookeeper.ZooKeeper)
[2019-05-07 07:30:56,260] INFO Client environment:host.name = mynode2 (org.apache.zookeeper.ZooKeeper)
[2019-05-07 07:30:56,261] INFO Client environment:java.version = 1.8.0_181 (org.apache.zookeeper.ZooKeeper)
```

```
165. [2019 - 05 - 07 07:30:56,261] INFO Client environment: java.
     vendor = Oracle Corporation (org.apache.zookeeper.ZooKeeper)
166. [2019 -05 -07 07:30:56,261] INFO Client environment:java.home
     = /software/jdk1.8.0_181/jre (org.apache.zookeeper.ZooKeeper)
167. [2019 - 05 - 07 07:30:56,261] INFO Client environment: java.
     class.path =:/software/kafka_2.11 - 0.11.0.3/bin/../libs/
     aopalliance - repackaged - 2.5.0 - b05.jar:/software/kafka_2.
     11 -0.11.0.3/bin/../libs/a
168. rgparse4j -0.7.0.jar:/software/kafka_2.11 -0.11.0.3/bin/../
     libs/commons - lang3 -
     3.5.jar:/software/kafka_2.11 -0.11.0.3/bin/../libs/connect -
     api -
     0.11.0.3.jar:/software/kafka_2.11 -0.11.0.3/bin/../libs/con-
     nect - file -
     0.11.0.3.jar:/software/kafka_2.11 -0.11.0.3/bin/../libs/con-
     nect - json -
     0.11.0.3.jar:/software/kafka_2.11 -0.11.0.3/bin/../libs/con-
     nect - runtime -
     0.11.0.3.jar:/software/kafka_2.11 -0.11.0.3/bin/../libs/con-
     nect - transforms -
     0.11.0.3.jar:/software/kafka_2.11 - 0.11.0.3/bin/../libs/
     guava -
     20.0.jar:/software/kafka_2.11 - 0.11.0.3/bin/../libs/hk2 -
     api -2.5.0 -
     b05.jar:/software/kafka_2.11 -0.11.0.3/bin/../libs/hk2 - lo-
     cator -2.5.0 -
     b05.jar:/software/kafka_2.11 - 0.11.0.3/bin/../libs/hk2 -
     utils -2.5.0 -
     b05.jar:/software/kafka_2.11 -0.11.0.3/bin/../libs/jackson -
     annotations -
     2.8.11.jar:/software/kafka_2.11 -0.11.0.3/bin/../libs/jack-
     son - core -
     2.8.10.jar:/software/kafka_2.11 -0.11.0.3/bin/../libs/jack-
     son - core -
     2.8.11.jar:/software/kafka_2.11 -0.11.0.3/bin/../libs/jack-
     son - databind -
     2.8.11.2.jar:/software/kafka_2.11 - 0.11.0.3/bin/../libs/
     jackson - jaxrs - base -
```

```
2.8.11.jar:/software/kafka_2.11-0.11.0.3/bin/../libs/jack-
son-jaxrs-json-provider-
2.8.11.jar:/software/kafka_2.11-0.11.0.3/bin/../libs/jack-
son-module-jaxb-annotations-
2.8.11.jar:/software/kafka_2.11-0.11.0.3/bin/../libs/javas-
sist-3.21.0-
GA.jar:/software/kafka_2.11-0.11.0.3/bin/../libs/javax.
annotation-api-
1.2.jar:/software/kafka_2.11-0.11.0.3/bin/../libs/javax.
inject-
1.jar:/software/kafka_2.11-0.11.0.3/bin/../libs/javax.
inject-2.5.0-
b05.jar:/software/kafka_2.11-0.11.0.3/bin/../libs/javax.
servlet-api-
3.1.0.jar:/software/kafka_2.11-0.11.0.3/bin/../libs/javax.
ws.rs-api-
2.0.1.jar:/software/kafka_2.11-0.11.0.3/bin/../libs/jersey-
client-
2.24.jar:/software/kafka_2.11-0.11.0.3/bin/../libs/jersey-
common-
2.24.jar:/software/kafka_2.11-0.11.0.3/bin/../libs/jersey-
container-servlet-
2.24.jar:/software/kafka_2.11-0.11.0.3/bin/../libs/jersey-
container-servlet-core-
2.24.jar:/software/kafka_2.11-0.11.0.3/bin/../libs/jersey-
guava-
2.24.jar:/software/kafka_2.11-0.11.0.3/bin/../libs/jersey-
media-jaxb-
2.24.jar:/software/kafka_2.11-0.11.0.3/bin/../libs/jersey-
server-
2.24.jar:/software/kafka_2.11-0.11.0.3/bin/../libs/jetty-
continuation-
9.2.22.v20170606.jar:/software/kafka_2.11-0.11.0.3/bin/../
libs/jetty-http-
9.2.22.v20170606.jar:/software/kafka_2.11-0.11.0.3/bin/../
libs/jetty-io-
9.2.22.v20170606.jar:/software/kafka_2.11-0.11.0.3/bin/../
libs/jetty-security-
```

```
9.2.22.v20170606.jar:/software/kafka_2.11-0.11.0.3/bin/../
libs/jetty-server-
9.2.22.v20170606.jar:/software/kafka_2.11-0.11.0.3/bin/../
libs/jetty-servlet-
9.2.22.v20170606.jar:/software/kafka_2.11-0.11.0.3/bin/../
libs/jetty-servlets-
9.2.22.v20170606.jar:/software/kafka_2.11-0.11.0.3/bin/../
libs/jetty-util-
9.2.22.v20170606.jar:/software/kafka_2.11-0.11.0.3/bin/../
libs/jopt-simple-
5.0.3.jar:/software/kafka_2.11-0.11.0.3/bin/../libs/kafka_
2.11-
0.11.0.3.jar:/software/kafka_2.11-0.11.0.3/bin/../libs/kafka_
2.11-0.11.0.3-
sources.jar:/software/kafka_2.11-0.11.0.3/bin/../libs/kafka_
2.11-0.11.0.3-test-
sources.jar:/software/kafka_2.11-0.11.0.3/bin/../libs/kafka-
clients-
0.11.0.3.jar:/software/kafka_2.11-0.11.0.3/bin/../libs/kafka-
log4j-appender-
0.11.0.3.jar:/software/kafka_2.11-0.11.0.3/bin/../libs/kafka-
streams-
0.11.0.3.jar:/software/kafka_2.11-0.11.0.3/bin/../libs/kafka-
streams-examples-
0.11.0.3.jar:/software/kafka_2.11-0.11.0.3/bin/../libs/kafka-
tools-
0.11.0.3.jar:/software/kafka_2.11-0.11.0.3/bin/../libs/
log4j-
1.2.17.jar:/software/kafka_2.11-0.11.0.3/bin/../libs/lz4-
1.3.0.jar:/software/kafka_2.11-0.11.0.3/bin/../libs/maven-
artifact-
3.5.0.jar:/software/kafka_2.11-0.11.0.3/bin/../libs/met-
rics-core-
2.2.0.jar:/software/kafka_2.11-0.11.0.3/bin/../libs/osgi-
resource-locator-
1.0.1.jar:/software/kafka_2.11-0.11.0.3/bin/../libs/plexus-
utils-
```

```
3.0.24.jar:/software/kafka_2.11-0.11.0.3/bin/../libs/re-
flections-
0.9.11.jar:/software/kafka_2.11-0.11.0.3/bin/../libs/rocks-
dbjni-
5.0.1.jar:/software/kafka_2.11-0.11.0.3/bin/../libs/scala-
library-
2.11.11.jar:/software/kafka_2.11-0.11.0.3/bin/../libs/sca-
la-parser-combinators_2.11-
1.0.4.jar:/software/kafka_2.11-0.11.0.3/bin/../libs/slf4j-
api-
1.7.25.jar:/software/kafka_2.11-0.11.0.3/bin/../libs/slf4j-
log4j12-
1.7.25.jar:/software/kafka_2.11-0.11.0.3/bin/../libs/snappy-
java-
1.1.2.6.jar:/software/kafka_2.11-0.11.0.3/bin/../libs/vali-
dation-api-
1.1.0.Final.jar:/software/kafka_2.11-0.11.0.3/bin/../libs/
zkclient-
0.10.jar:/software/kafka_2.11-0.11.0.3/bin/../libs/zookeeper-
3.4.10.jar
(org.apache.zookeeper.ZooKeeper)[2019-05-07 07:30:56,261]
INFO Client
environment:java.library.path=/usr/java/packages/lib/
amd64:/usr/lib64:/lib64:/lib:/usr/lib (org.apache.zookeeper.
ZooKeeper)
169.[2019-05-07 07:30:56,261] INFO Client environment:java.io.
tmpdir=/tmp (org.apache.zookeeper.ZooKeeper)
170.[2019-05-07 07:30:56,261] INFO Client environment:java.com-
piler=<NA> (org.apache.zookeeper.ZooKeeper)
171.[2019-05-07 07:30:56,261] INFO Client environment:os.name=
Linux (org.apache.zookeeper.ZooKeeper)
172.[2019-05-07 07:30:56,261] INFO Client environment:os.arch=
amd64 (org.apache.zookeeper.ZooKeeper)
173.[2019-05-07 07:30:56,261] INFO Client environment:os.version
=2.6.32-504.el6.x86_64 (org.apache.zookeeper.ZooKeeper)
174.[2019-05-07 07:30:56,261] INFO Client environment:user.name
=root (org.apache.zookeeper.ZooKeeper)
```

```
[2019-05-07 07:30:56,261] INFO Client environment:user.home=/root (org.apache.zookeeper.ZooKeeper)
[2019-05-07 07:30:56,261] INFO Client environment:user.dir=/software/kafka_2.11-0.11.0.3/bin (org.apache.zookeeper.ZooKeeper)
... ...
... ...
[2019-05-07 07:30:59,012] INFO Truncating flumetopic-2 to 3 has no effect as the largest offset in the log is 2. (kafka.log.Log)
[2019-05-07 07:30:59,012] INFO Truncating mytest-0 to 21 has no effect as the largest offset in the log is 20. (kafka.log.Log)
[2019-05-07 07:30:59,012] INFO Truncating flumetopic-1 to 2 has no effect as the largest offset in the log is 1. (kafka.log.Log)
[2019-05-07 07:30:59,012] INFO Truncating mytest-2 to 21 has no effect as the largest offset in the log is 20. (kafka.log.Log)
[2019-05-07 07:30:59,012] INFO Truncating flumetopic-0 to 2 has no effect as the largest offset in the log is 1. (kafka.log.Log)
[2019-05-07 07:30:59,013] INFO [Group Metadata Manager on Broker 1]: Finished loading offsets and group metadata from __consumer_offsets-2 in 23 milliseconds. (kafka.coordinator.group.GroupMetadataManager)
[2019-05-07 07:30:59,018] INFO [Group Metadata Manager on Broker 1]: Finished loading offsets and group metadata from __consumer_offsets-5 in 0 milliseconds. (kafka.coordinator.group.GroupMetadataManager)
[2019-05-07 07:30:59,043] INFO [ReplicaFetcherManager on broker 1] Removed fetcher for partitions __consumer_offsets-8,mytopic-0,__consumer_offsets-35,__consumer_offsets-41,__consumer_offsets-23,__consumer_offsets-47,__consumer_offsets-38,__consumer_offsets-17,__consumer_offsets-11,__consumer_offsets-2,__consumer_offsets-14,__consumer_offsets-20,__consumer_offsets-44,__consumer_offsets-5,__consumer_offsets-26,__consumer_offsets-29,__consumer_offsets-32 (kafka.server.ReplicaFetcherManager)
[2019-05-07 07:30:59,043] INFO Partition [mytopic,0] on broker 1: mytopic-0 starts at Leader Epoch 4 from offset 1. Previous Leader Epoch was: 3 (kafka.cluster.Partition)
```

```
188.[2019-05-07 07:30:59,055] INFO Partition [__consumer_offsets,29] on broker 1: __consumer_offsets-29 starts at Leader Epoch 24 from offset 244. Previous Leader Epoch was: 23 (kafka.cluster.Partition)
189.[2019-05-07 07:30:59,057] INFO Partition [__consumer_offsets,26] on broker 1: __consumer_offsets-26 starts at Leader Epoch 24 from offset 2637. Previous Leader Epoch was: 23 (kafka.cluster.Partition)
190.[2019-05-07 07:30:59,057] INFO Partition [__consumer_offsets,23] on broker 1: __consumer_offsets-23 starts at Leader Epoch 24 from offset 185. Previous Leader Epoch was: 23 (kafka.cluster.Partition)
191.[2019-05-07 07:30:59,058] INFO Partition [__consumer_offsets,20] on broker 1: __consumer_offsets-20 starts at Leader Epoch 24 from offset 0. Previous Leader Epoch was: 23 (kafka.cluster.Partition)
192.[2019-05-07 07:30:59,058] INFO Partition [__consumer_offsets,17] on broker 1: __consumer_offsets-17 starts at Leader Epoch 24 from offset 1822. Previous Leader Epoch was: 23 (kafka.cluster.Partition)
193.[2019-05-07 07:30:59,058] INFO Partition [__consumer_offsets,14] on broker 1: __consumer_offsets-14 starts at Leader Epoch 24 from offset 0. Previous Leader Epoch was: 23 (kafka.cluster.Partition)
194.[2019-05-07 07:30:59,059] INFO Partition [__consumer_offsets,11] on broker 1: __consumer_offsets-11 starts at Leader Epoch 24 from offset 0. Previous Leader Epoch was: 23 (kafka.cluster.Partition)
195.[2019-05-07 07:30:59,060] INFO Partition [__consumer_offsets,8] on broker 1: __consumer_offsets-8 starts at Leader Epoch 24 from offset 97. Previous Leader Epoch was: 23 (kafka.cluster.Partition)
196.[2019-05-07 07:30:59,060] INFO Partition [__consumer_offsets,5] on broker 1: __consumer_offsets-5 starts at Leader Epoch 24 from offset 0. Previous Leader Epoch was: 23 (kafka.cluster.Partition)
197.[2019-05-07 07:30:59,060] INFO Partition [__consumer_offsets,
```

```
    2] on broker 1: __consumer_offsets -2 starts at Leader Epoch 24
    from offset 0. Previous Leader Epoch was: 23 (kafka.cluster.
    Partition)
198. [2019-05-07 07:30:59,060] INFO Partition [__consumer_off-
    sets,47] on broker 1: __consumer_offsets -47 starts at Leader
    Epoch 24 from offset 714. Previous Leader Epoch was: 23 (kafka.
    cluster.Partition)
199. [2019-05-07 07:30:59,060] INFO Partition [__consumer_off-
    sets,38] on broker 1: __consumer_offsets -38 starts at Leader
    Epoch 24 from offset 66. Previous Leader Epoch was: 23 (kafka.
    cluster.Partition)
200. [2019-05-07 07:30:59,061] INFO Partition [__consumer_off-
    sets,35] on broker 1: __consumer_offsets -35 starts at Leader
    Epoch 24 from offset 0. Previous Leader Epoch was: 23 (kafka.
    cluster.Partition)
201. [2019-05-07 07:30:59,061] INFO Partition [__consumer_off-
    sets,44] on broker 1: __consumer_offsets -44 starts at Leader
    Epoch 24 from offset 14322. Previous Leader Epoch was: 23 (kaf-
    ka.clusterPartition)
202. [2019-05-07 07:30:59,061] INFO Partition [__consumer_off-
    sets,32] on broker 1: __consumer_offsets -32 starts at Leader
    Epoch 24 from offset 0. Previous Leader Epoch was: 23 (kafka.
    cluster.Partition)
203. [2019-05-07 07:30:59,061] INFO Partition [__consumer_off-
    sets,41] on broker 1: __consumer_offsets -41 starts at Leader
    Epoch 24 from offset 332. Previous Leader Epoch was: 23 (kafka.
    cluster.Partition)
204. [2019-05-07 07:30:59,110] INFO [Group Metadata Manager on
    Broker 1]: Finished loading offsets and group metadata from __
    consumer_offsets-8 in 92 milliseconds. (kafka.coordinator.
    group.GroupMetadataManager)
205. [2019-05-07 07:30:59,111] INFO [Group Metadata Manager on
    Broker 1]: Finished loading offsets and group metadata from __
    consumer_offsets-11 in 0 milliseconds. (kafka.coordinator.
    group.GroupMetadataManager)
206. [2019-05-07 07:30:59,111] INFO [Group Metadata Manager on
    Broker 1]: Finished loading offsets and group metadata from __
```

```
consumer_offsets-14 in 0 milliseconds. (kafka.coordinator.group.GroupMetadataManager)
[2019-05-07 07:30:59,177] INFO [Group Metadata Manager on Broker 1]: Finished loading offsets and group metadata from __consumer_offsets-17 in 65 milliseconds. (kafka.coordinator.group.GroupMetadataManager)
[2019-05-07 07:30:59,178] INFO [Group Metadata Manager on Broker 1]: Finished loading offsets and group metadata from __consumer_offsets-20 in 0 milliseconds. (kafka.coordinator.group.GroupMetadataManager)
[2019-05-07 07:30:59,211] INFO [Group Metadata Manager on Broker 1]: Finished loading offsets and group metadata from __consumer_offsets-23 in 33 milliseconds. (kafka.coordinator.group.GroupMetadataManager)
[2019-05-07 07:30:59,244] INFO [Group Metadata Manager on Broker 1]: Finished loading offsets and group metadata from __consumer_offsets-26 in 33 milliseconds. (kafka.coordinator.group.GroupMetadataManager)
[2019-05-07 07:30:59,278] INFO [Group Metadata Manager on Broker 1]: Finished loading offsets and group metadata from __consumer_offsets-29 in 34 milliseconds. (kafka.coordinator.group.GroupMetadataManager)
[2019-05-07 07:30:59,278] INFO [Group Metadata Manager on Broker 1]: Finished loading offsets and group metadata from __consumer_offsets-32 in 0 milliseconds. (kafka.coordinator.group.GroupMetadataManager)
[2019-05-07 07:30:59,278] INFO [Group Metadata Manager on Broker 1]: Finished loading offsets and group metadata from __consumer_offsets-35 in 0 milliseconds. (kafka.coordinator.group.GroupMetadataManager)
[2019-05-07 07:30:59,310] INFO [Group Metadata Manager on Broker 1]: Finished loading offsets and group metadata from __consumer_offsets-38 in 32 milliseconds. (kafka.coordinator.group.GroupMetadataManager)
[2019-05-07 07:30:59,344] INFO [Group Metadata Manager on Broker 1]: Finished loading offsets and group metadata from __consumer_offsets-41 in 34 milliseconds. (kafka.coordinator.group.GroupMetadataManager)
```

```
[2019-05-07 07:30:59,414] INFO [Group Metadata Manager on Broker 1]: Finished loading offsets and group metadata from __consumer_offsets-44 in 70 milliseconds. (kafka.coordinator.group.GroupMetadataManager)
[2019-05-07 07:30:59,448] INFO [Group Metadata Manager on Broker 1]: Finished loading offsets and group metadata from __consumer_offsets-47 in 34 milliseconds. (kafka.coordinator.group.GroupMetadataManager)
......
......
```

在 mynode1、mynode2、mynode3 节点上启动 Kafka 集群的命令执行后，Kafka 进程会占用当前节点的前台终端。为了避免这个问题，在前面项目中提到在 mynode1、mynode2、mynode3 每台节点路径/software/kafka_2.11-0.11.0.3/下配置 startkafka.sh 脚本。脚本内容如下：

```
nohup ./bin/kafka-server-start.sh ./config/server.properties > kafka.log 2 >&1 &
```

上面脚本的意思是将 Kafka 启动后，以后台方式启动，不占用当前节点的终端。将启动产生的日志写入文件 kafka.log 中，此后，再涉及启动 Kafka 时，就可以启动当前脚本 startkafka.sh，这样就避免了 Kafka 进程占用节点终端的问题。

下面执行查看 Kafka 集群中的 Topic 命令，执行命令如下：

```
[root@mynode1 ~]# cd /software/kafka_2.11-0.11.0.3/bin/
[root@mynode1 bin]# ./kafka-topics.sh --zookeeper mynode3:2181,mynode4:2181,mynode5:2181 --list
```

上述命令可以在 Kafka 集群的任何一台 Broker 节点中执行，以上命令参数的解释如下：

- zookeeper：指定当前 Kafka 集群的 ZooKeeper 节点，便于寻找 Topic 元数据信息。
- list：列举当前 Kafka 集群中所有的 Topic 信息。

由于刚搭建好 Kafka 集群，还没有创建 Topic 信息，所以目前查询不到 Topic 的详细信息，现在介绍如何使用命令在 Kafka 中创建 Topic。

### 6.2.2 创建 Topic

Kafka 消息系统中的 Topic 主要是存储不同类别的消息。在 Kafka 中创建 Topic 时，在任何一台 Broker 节点执行创建命令都可以，这里以在 mynode1 节点执行创建命令为例，创建 Kafka Topic 的命令如下：

```
[root@mynode1 ~]# cd /software/kafka_2.11-0.11.0.3/bin/
[root@mynode1 bin]# ./kafka-topics.sh --zookeeper mynode3:
```

```
   2181,mynode4:2181,mynode5:2181 --create --topic firsttopic --
   partitions 3 --replication-factor 3
3. Created topic "firsttopic".
```

上面命令是在 Kafka 消息系统中创建了名称为 firsttopic 的消息队列。关于上面命令的详细解释如下。

kafka-topic. sh：执行 Kafka Topic 脚本，这个脚本可以创建、修改、删除、查看 Kafka 集群中的信息。

- create：执行当前操作，创建 Topic。
- topic：指定当前创建 Topic 的名称。
- partitions：指定当前 Topic 的分区数，方便生产者生产消息和消费者并行消费消息。
- replication-factor：指定当前 Topic 的分区数据的副本个数，这个个数不能超过当前 Kafka 集群所有的 Broker 节点。

通过以上命令创建 Topic 后，出现“Created topic "firsttopic".”，代表创建 Topic 成功。下面使用命令验证 Topic 是否创建成功，在任意一台 Broker 节点上启动执行查询命令即可查询，这里还是在 mynode1 节点上执行查询命令，命令如下：

```
1. [root@mynode1 ~]# cd /software/kafka_2.11-0.11.0.3/bin/
2. [root@mynode1 bin]# ./kafka-topics.sh --zookeeper mynode3:
   2181,mynode4:2181,mynode5:2181 --list
3. firsttopic
```

通过以上查询命令的验证，“firsttopic”创建成功。在××系统中，同样可以创建收集各个网站数据的 Topic，按照以上方式创建 Topic，为将从××系统采集的数据存入 Kafka 中奠定了基础。

### 6.2.3　向 Topic 生产消息

Topic 创建完成之后，测试向当前“firsttopic”消息队列中生产消息。向 Kafka 中的 Topic 中生产消息的生产者可以多种多样，这些生产者可以是分布式采集工具 Flume、Kafka Console 生产者、Java 或者 Scala 调用生产者 API 所形成的客户端等。无论是以上哪种生产者，都可以生成各种各样的消息输入 Kafka 集群中。Kafka 集群的吞吐量可以达到百兆每秒的速度，所以 Kafka 的生产者也可以是多生产者并行生产的系统。

下面将 Kafka Console 当作消息的生产者，向当前“firsttopic”消息队列中生产消息。可以在 Kafka 任意一台 Broker 节点上执行向 Topic 中生产消息的命令，这里以在 mynode1 节点执行生产命令为例，在 mynode1 上执行如下生产命令：

```
1. [root@mynode1 ~]# cd /software/kafka_2.11-0.11.0.3/bin/
2. [root@mynode1 bin]# ./kafka-console-producer.sh --broker-list
   mynode1:9092,mynode2:9092,mynode3:9092 --topic firsttopic
3. >
```

对以上执行命令的参数解释如下：

- kafka-console-producer.sh：Console 的生产的脚本。
- --broker-list：Kafka 集群的 Broker 节点，这里是执行 Kafka 生产者将消息生产到哪个 Kafka 集群中。
- --topic：执行生产者生产消息的 Topic，也就是将消息写入哪个 Topic 中。

以上命令执行完成之后，Kafka Console 就处于等待输入消息的状态。向 Kafka 的 "first-topic" 中输入如下消息：

```
1. >11111
2. >22222
3. >33333
4. >44444
5. >55555
6. >
```

通过以上输入命令，向 Kafka 的 "firsttopic" 中输入了 5 条消息，下面检验这 5 条消息是否全部成功生产到 Kafka 的 "firsttopic" 中。

### 6.2.4 从 Topic 消费消息

要检验消息有没有被成功写入 Kafka 的 "firsttopic" 中，可以通过消费 "firsttopic" 中的数据来完成。Kafka 的消费者可以是 Java 或者 Scala 的 API 消费者，也可以是 SparkStreaming 计算系统，还可以是 Kafka Console 消费者。无论是以上哪种消费者，都可以读取到成功写入 "firsttopic" 中的数据，这里将 Kafka Console 当作消费者来消费 "firsttopoic" 中的数据，查看是否成功地将消息生产到了消息队列中。

将 Kafka Console 当作 Kafka Topic 的消费者时，可以在 Kafka 集群中的任意一台 Broker 节点启动 Console 控制台消费者，这里以在 mynode1 中启动 Kafka Console 消费者为例，执行如下命令：

```
1. [root@mynode1 ~]# cd /software/kafka_2.11-0.11.0.3/bin/
2. [root@mynode1 bin]# ./kafka-console-consumer.sh --bootstrap-
   server mynode1:9092,mynode2:9092,mynode3:9092  --topic first-
   topic
```

对以上命令参数的解释如下：

- kafka-console-consumer.sh：消息消费者的执行脚本。
- --bootstrap-server：指定 Kafka 集群的 Broker 节点。指定这个参数的原因是指定当前消费者消费消息的位置信息由 Kafka 集群管理。
- --topic：指定当前消费消息数据来源的 Topic。

输入以上命令后，没有发现消费之前输入的 5 条消息，原因是 Kafka 在启动Console 消费者之后，默认消费 Kafka Topic 的位置为最新位置。也就是说，当启动 Kafka Console 消费者之后，只能消费到之后输入的消息数据。

这里也可以设置参数，用于控制 Kafka Console 消费者启动之后从头消费当前 Kafka 的

“firsttopic”中的消息，执行如下命令：

```
[root@mynode1 bin]# ./kafka-console-consumer.sh --bootstrap-server mynode1:9092,mynode2:9092,mynode3:9092 --topic first-topic --from-beginning
22222
55555
33333
11111
44444
```

执行以上命令之后，能够发现消费到了之前向“firsttopic”中生产的5条消息。以上执行的命令中，参数“--from-beginning”的意思是指定当前Kafka Console消费Kafka中的消息时从头开始消费。

此时经历了创建Topic、向Topic中生产消息、消费Topic中的消息几个过程，完整的Kafka消息队列的创建、生产、消费流程基本测试完成。这样，在××系统中可以使用当前Kafka消息系统来缓存收集到的网站数据。

### 6.2.5 offset查看

以上几个小节中，在Kafka消息队列中创建了“firsttopic”消息队列，同时，向当前这个消息队列中生产消息，以及消费者当前消息队列中的数据。在Kafka中，当前Kafka Console消费者消费消息的位置数据信息存储在Kafka中，在Kafka的旧版本中，例如在Kafka 0.8.2版本中，消费者的消费数据位置信息是存储在ZooKeeper中的。但是在Kafka 0.8.2版本之后，消费者的消费数据位置信息默认由Kafka集群管理。这是因为ZooKeeper不擅长频繁读写，而维护消费者消费消息的位置信息需要频繁地读写。

在新版本中，Kafka是以消费者组为单位将消费者消费的数据位置信息存储在一个名为“__consumer_offsets”的Topic中的，下面使用命令查看当前Kafka集群中Kafka Console消费者消费消息的位置。

查看Kafka Console消费者消费消息的位置时，由于保存在Kafka中的消费者消费位置都是以消费者组为单位的，所以，这里先查询有哪些消费者组。

在任意一台Kafka的Broker节点中执行如下命令来查询消费者组。

```
[root@mynode2 kafka_2.11-0.11.0.3]# cd /software/kafka_2.11-0.11.0.3/bin/
[root@mynode2 bin]# ./kafka-consumer-groups.sh --bootstrap-server mynode1:9092,mynode2:9092,mynode3:9092 --list
Note: This will only show information about consumers that use the Java consumer API (non-ZooKeeper-based consumers).

console-consumer-82938
```

对以上参数的解释如下：

- kafka - console - group. sh：查询消费者组的执行脚本。
- -- bootstrap - server：执行查询的 Kafka 集群节点。
- -- list：列出当前 Kafka 集群中所有的消费者组。

使用以上命令查询出一个名称为“console - consumer - 82938”的消费者组。在 Kafka 集群中，之前只启动过一个消费者，Kafka 集群会为当前的消费自动分配一个消费者组名称，“console - consumer - 82938”正是 Kafka 集群为 Kafka Console 分配的消费者组名称。

使用如下命令查询当前组下消费 Kafka 消息队列的位置信息：

```
1. [root@mynode2 kafka_2.11 -0.11.0.3]# cd /software/kafka_2.11 -
   0.11.0.3/bin/
2. [root@mynode2 bin]# ./kafka - consumer - groups.sh --bootstrap -
   server mynode1:9092,mynode2:9092,mynode3:9092 --describe --
   group console - consumer - 82938
3. Note: This will only show information about consumers that use
   the Java consumer API (non - ZooKeeper - based consumers).
4.
5. Consumer group 'console - consumer - 82938' has no active members.
6.
7. TOPIC PARTITION CURRENT - OFFSET LOG - END - OFFSET LAG CONSUMER -
   ID HOST CLIENT - ID
8. firsttopic 1 2 2 0 - - -
9. firsttopic 0 2 2 0 - - -
10. firsttopic 2 1 1 0 - -
```

对以上查询命令的参数解释如下：

- kafka - consumer - groups. sh：查询 Kafka 消费者组的执行命令。
- -- bootstrap - server：执行 Kafka 集群。
- -- describe：指定当前是需要查看 Kakfa 消费者组的描述信息。
- -- group：指定查看的消费者组的名称。

在以上命令查询出的信息中，各个列代表的意思如下：

- TOPIC：消费的当前 Topic 名称。
- PARTITION：消费的当前 Topic 的 Partiton。
- CURRENT - OFFSET：当前消费者消费的位置信息量。
- LOG - END - OFFSET：当前生产者生产数的位置信息量。
- LAG：当前消费者没有消费的消息信息量。
- CONSUMER - ID：当前消费者的 ID。
- HOST：当前主机节点。

- CLIENT－ID：客户端 ID。

通过以上命令，发现当前“console－consumer－82938”消费者组已经成功消费了 Kafka 消息队列“firsttopic”中的数据。

### 6.2.6 删除 Topic 信息

在 Kafka 集群中，有时不再需要某个消息队列来保存消息，需要将某个消息队列删除，可以使用命令将 Kafka 中的某些消息队列删除。

假设需要将 Kafka 中的消息队列“firsttopic”删除，可以在 Kafka 任意的 Broker 节点执行删除 Kafka 消息队列的命令。这里以在 mynode1 节点删除“firsttopic”为例。

在 mynode1 中执行如下删除命令：

```
[root@mynode1 ~]# cd /software/kafka_2.11-0.11.0.3/bin/
[root@mynode1 bin]# ./kafka-topics.sh --zookeeper mynode3:2181,mynode4:2181,mynode5:2181 --delete --topic firsttopic
Topic firsttopic is marked for deletion.
Note: This will have no impact if delete.topic.enable is not set to true.
-
```

对以上执行命令的参数解释如下：

kafka－topic. sh：关于 Kafka Topic 的执行脚本。

－－zookeeper：指定协调 Kafka 集群的 ZooKeeper 集群。

－－delete：指定删除 Kafka 的命令。

－－topic：指定要删除的 Topic 名称。

执行完以上命令之后，出现“Note：This will have no impact if delete. topic. enable is not set to true.”提示。这个提示的意思是本条删除命令如果没有设置“delete. topic. enable”参数为 true，则当前执行的命令无效。在 Kafka 集群中，“delete. topic. enable”默认为 true。

执行如下命令查询当前 Kafka 集群中的“firsttopic”是否删除成功：

```
[root@mynode1 ~]# cd /software/kafka_2.11-0.11.0.3/bin/
[root@mynode1 bin]# ./kafka-topics.sh --zookeeper mynode3:2181,mynode4:2181,mynode5:2181 --list
__consumer_offsets
firsttopic
```

对以上参数的解释如下：

- kafka－topic. sh：Kafka 集群的 Topic 的查询执行脚本。
- －－zookeeper：指定协调当前 Kafka 集群的 ZooKeeper 集群。
- －－list：查看当前 Kafka 集群的 Topic 信息。

通过以上命令查询，发现“firsttopic”确实没有被删除，下面在 Kafka 中配置“delete.topic.enable”为 true，开启删除 Topic 的功能。

配置步骤如下：

①停止 Kafka 集群。

由于之前 Kafka 集群一直启动，现在首先停止 Kafka 集群。停止 Kafka 集群有单独的停止脚本，这个脚本位于 Kafka Broker 节点的路径/bin 下，名称为“kafka-server-stop.sh”，目前这个脚本存在一些问题，需要修改脚本后再停止每个 Kafka 集群。

下面在 Kafka 每台 Broker 节点 mynode1、mynode2、mynode3 上修改 Kafka 路径中的“kafka-server-stop.sh”脚本，完整的脚本内容如下：

```
#!/bin/sh
# Licensed to the Apache Software Foundation (ASF) under one or more
# contributor license agreements. See the NOTICE file distributed with
# this work for additional information regarding copyright ownership.
# The ASF licenses this file to You under the Apache License, Version 2.0
# (the "License"); you may not use this file except in compliance with
# the License. You may obtain a copy of the License at
#
# http://www.apache.org/licenses/LICENSE-2.0
#
# Unless required by applicable law or agreed to in writing, software
# distributed under the License is distributed on an "AS IS" BASIS,
# WITHOUT WARRANTIES OR CONDITIONS OF ANY KIND, either express or implied.
# See the License for the specific language governing permissions and
# limitations under the License.
PIDS=$(ps ax | grep -i 'Kafka' | grep java | grep -v grep | awk '{print $1}')

if [ -z "$PIDS" ]; then
echo "No kafka server to stop"
```

```
exit 1
else
kill -s TERM $PIDS
fi
```

以上脚本需要在每台 Kafka Broker 节点上修改，修改成功之后再执行如下停止 Kafka 集群的命令：

```
#在 mynode1 节点上执行如下命令
[root@mynode1 ~]# cd /software/kafka_2.11-0.11.0.3/bin/
[root@mynode1 bin]# ./kafka-server-stop.sh
#检查是否成功将 Kafaka 进程停止,在 mynode1 上执行如下命令
[root@mynode1 bin]#jps
5322 Jps
#在 mynode2 节点上执行如下命令
[root@mynode2 ~]# cd /software/kafka_2.11-0.11.0.3/bin/
[root@mynode2 bin]# ./kafka-server-stop.sh
#检查是否成功将 Kafaka 进程停止,在 mynode2 上执行如下命令
[root@mynode2 bin]#jps
1298 Jps
#在 mynode3 节点上执行如下命令
[root@mynode3 ~]# cd /software/kafka_2.11-0.11.0.3/bin/
[root@mynode3 bin]# ./kafka-server-stop.sh
#检查是否成功将 Kafaka 进程停止,在 mynode1=3 上执行如下命令
[root@mynode3 bin]#jps
3647 Jps
```

经过以上命令的执行，Kafka 集群成功停止。

②在每台 Kafka 集群的 Broker 节点配置参数。

下面需要在每台 Kafka 集群的 Broker 节点配置“delete. topic. enable”为 true，在每台 Kafka Broker 节点路径“/software/kafka_ 2. 11 -0. 11. 0. 3/config”下配置 server. properties 文件，在这个文件后面追加如下内容：

```
delete.topic.enable=true
```

③重启 Kafka 集群。

在 mynode1 上单独启动 Kafka，执行如下命令：

```
[root@mynode1 ~]# cd /software/kafka_2.11-0.11.0.3/bin/
[root@mynode1 bin]# ./kafka-server-start.sh /software/kafka_2.11-0.11.0.3/config/server.properties
[2019-05-10 21:38:17,056] INFO KafkaConfig values:
advertised.host.name = null
advertised.listeners = null
advertised.port = null
alter.config.policy.class.name = null
authorizer.class.name =
auto.create.topics.enable = true
auto.leader.rebalance.enable = true
background.threads = 10
broker.id = 0
broker.id.generation.enable = true
broker.rack = null
compression.type = producer
connections.max.idle.ms = 600000
controlled.shutdown.enable = true
controlled.shutdown.max.retries = 3
controlled.shutdown.retry.backoff.ms = 5000
controller.socket.timeout.ms = 30000
create.topic.policy.class.name = null
default.replication.factor = 1
delete.records.purgatory.purge.interval.requests = 1
delete.topic.enable = true
fetch.purgatory.purge.interval.requests = 1000
group.initial.rebalance.delay.ms = 0
group.max.session.timeout.ms = 300000
group.min.session.timeout.ms = 6000
host.name =
inter.broker.listener.name = null
inter.broker.protocol.version = 0.11.0-IV2
leader.imbalance.check.interval.seconds = 300
leader.imbalance.per.broker.percentage = 10
listener.security.protocol.map = SSL:SSL,SASL_PLAINTEXT:SASL_PLAINTEXT,TRACE:TRACE,SASL_SSL:SASL_SSL,PLAINTEXT:PLAINTEXT
listeners = null
```

```
log.cleaner.backoff.ms = 15000
log.cleaner.dedupe.buffer.size = 134217728
log.cleaner.delete.retention.ms = 86400000
log.cleaner.enable = true
log.cleaner.io.buffer.load.factor = 0.9
log.cleaner.io.buffer.size = 524288
log.cleaner.io.max.bytes.per.second = 1.7976931348623157E308
log.cleaner.min.cleanable.ratio = 0.5
log.cleaner.min.compaction.lag.ms = 0
log.cleaner.threads = 1
log.cleanup.policy = [delete]
log.dir = /tmp/kafka-logs
log.dirs = /kafka-logs
log.flush.interval.messages = 9223372036854775807
log.flush.interval.ms = null
log.flush.offset.checkpoint.interval.ms = 60000
log.flush.scheduler.interval.ms = 9223372036854775807
log.flush.start.offset.checkpoint.interval.ms = 60000
log.index.interval.bytes = 4096
log.index.size.max.bytes = 10485760
log.message.format.version = 0.11.0-IV2
log.message.timestamp.difference.max.ms = 9223372036854775807
log.message.timestamp.type = CreateTime
log.preallocate = false
log.retention.bytes = -1
log.retention.check.interval.ms = 300000
log.retention.hours = 168
log.retention.minutes = null
log.retention.ms = null
log.roll.hours = 168
log.roll.jitter.hours = 0
log.roll.jitter.ms = null
log.roll.ms = null
log.segment.bytes = 1073741824
log.segment.delete.delay.ms = 60000
max.connections.per.ip = 2147483647
max.connections.per.ip.overrides =
message.max.bytes = 1000012
```

```
74. metric.reporters = []
75. metrics.num.samples = 2
76. metrics.recording.level = INFO
77. metrics.sample.window.ms = 30000
78. min.insync.replicas = 1
79. num.io.threads = 8
80. num.network.threads = 3
81. num.partitions = 1
82. num.recovery.threads.per.data.dir = 1
83. num.replica.fetchers = 1
84. offset.metadata.max.bytes = 4096
85. offsets.commit.required.acks = -1
86. offsets.commit.timeout.ms = 5000
87. offsets.load.buffer.size = 5242880
88. offsets.retention.check.interval.ms = 600000
89. offsets.retention.minutes = 1440
90. offsets.topic.compression.codec = 0
91. offsets.topic.num.partitions = 50
92. offsets.topic.replication.factor = 1
93. offsets.topic.segment.bytes = 104857600
94. port = 9092
95. principal.builder.class = class org.apache.kafka.common.secur-
    ity.auth.DefaultPrincipalBuilder
96. producer.purgatory.purge.interval.requests = 1000
97. queued.max.requests = 500
98. quota.consumer.default = 9223372036854775807
99. quota.producer.default = 9223372036854775807
100. quota.window.num = 11
101. quota.window.size.seconds = 1
102. replica.fetch.backoff.ms = 1000
103. replica.fetch.max.bytes = 1048576
104. replica.fetch.min.bytes = 1
105. replica.fetch.response.max.bytes = 10485760
106. replica.fetch.wait.max.ms = 500
107. replica.high.watermark.checkpoint.interval.ms = 5000
108. replica.lag.time.max.ms = 10000
109. replica.socket.receive.buffer.bytes = 65536
110. replica.socket.timeout.ms = 30000
```

```
replication.quota.window.num = 11
replication.quota.window.size.seconds = 1
request.timeout.ms = 30000
reserved.broker.max.id = 1000
sasl.enabled.mechanisms = [GSSAPI]
sasl.kerberos.kinit.cmd = /usr/bin/kinit
sasl.kerberos.min.time.before.relogin = 60000
sasl.kerberos.principal.to.local.rules = [DEFAULT]
sasl.kerberos.service.name = null
sasl.kerberos.ticket.renew.jitter = 0.05
sasl.kerberos.ticket.renew.window.factor = 0.8
sasl.mechanism.inter.broker.protocol = GSSAPI
security.inter.broker.protocol = PLAINTEXT
socket.receive.buffer.bytes = 102400
socket.request.max.bytes = 104857600
socket.send.buffer.bytes = 102400
ssl.cipher.suites = null
ssl.client.auth = none
ssl.enabled.protocols = [TLSv1.2, TLSv1.1, TLSv1]
ssl.endpoint.identification.algorithm = null
ssl.key.password = null
ssl.keymanager.algorithm = SunX509
ssl.keystore.location = null
ssl.keystore.password = null
ssl.keystore.type = JKS
ssl.protocol = TLS
ssl.provider = null
ssl.secure.random.implementation = null
ssl.trustmanager.algorithm = PKIX
ssl.truststore.location = null
ssl.truststore.password = null
ssl.truststore.type = JKS
transaction.abort.timed.out.transaction.cleanup.interval.ms = 60000
transaction.max.timeout.ms = 900000
transaction.remove.expired.transaction.cleanup.interval.ms = 3600000
transaction.state.log.load.buffer.size = 5242880
```

```
transaction.state.log.min.isr=1
transaction.state.log.num.partitions=50
transaction.state.log.replication.factor=1
transaction.state.log.segment.bytes=104857600
transactional.id.expiration.ms=604800000
unclean.leader.election.enable=false
zookeeper.connect=mynode3:2181,mynode4:2181,mynode5:2181
zookeeper.connection.timeout.ms=6000
zookeeper.session.timeout.ms=6000
zookeeper.set.acl=false
zookeeper.sync.time.ms=2000
(kafka.server.KafkaConfig)
[2019-05-10 21:38:17,096] INFO starting (kafka.server.KafkaServer)
[2019-05-10 21:38:17,097] INFO Connecting to zookeeper on mynode3:2181,mynode4:2181,mynode5:2181 (kafka.server.KafkaServer)
[2019-05-10 21:38:17,120] INFO Starting ZkClient event thread. (org.I0Itec.zkclient.ZkEventThread)
[2019-05-10 21:38:17,125] INFO Client environment:zookeeper.version=3.4.10-39d3a4f269333c922ed3db283be479f9deacaa0f, built on 03/23/2017 10:13 GMT (org.apache.zookeeper.ZooKeeper)
[2019-05-10 21:38:17,125] INFO Client environment:host.name=mynode1 (org.apache.zookeeper.ZooKeeper)
[2019-05-10 21:38:17,125] INFO Client environment:java.version=1.8.0_181 (org.apache.zookeeper.ZooKeeper)
[2019-05-10 21:38:17,125] INFO Client environment:java.vendor=Oracle Corporation (org.apache.zookeeper.ZooKeeper)
[2019-05-10 21:38:17,125] INFO Client environment:java.home=/software/jdk1.8.0_181/jre (org.apache.zookeeper.ZooKeeper)
[2019-05-10 21:38:17,125] INFO Client environment:java.class.path=:/software/kafka_2.11-0.11.0.3/bin/../libs/aopalliance-repackaged-2.5.0-b05.jar:/software/kafka_2.11-0.11.0.3/bin/../libs/a
rgparse4j-0.7.0.jar:/software/kafka_2.11-0.11.0.3/bin/../libs/commons-lang3-
3.5.jar:/software/kafka_2.11-0.11.0.3/bin/../libs/connect-api-
```

```
0.11.0.3.jar:/software/kafka_2.11-0.11.0.3/bin/../libs/con-
nect-file-
0.11.0.3.jar:/software/kafka_2.11-0.11.0.3/bin/../libs/con-
nect-json-
0.11.0.3.jar:/software/kafka_2.11-0.11.0.3/bin/../libs/con-
nect-runtime-
0.11.0.3.jar:/software/kafka_2.11-0.11.0.3/bin/../libs/con-
nect-transforms-
0.11.0.3.jar:/software/kafka_2.11-0.11.0.3/bin/../libs/
guava-
20.0.jar:/software/kafka_2.11-0.11.0.3/bin/../libs/hk2-
api-2.5.0-
b05.jar:/software/kafka_2.11-0.11.0.3/bin/../libs/hk2-
locator-2.5.0-
b05.jar:/software/kafka_2.11-0.11.0.3/bin/../libs/hk2-
utils-2.5.0-
b05.jar:/software/kafka_2.11-0.11.0.3/bin/../libs/jackson
-annotations-
2.8.11.jar:/software/kafka_2.11-0.11.0.3/bin/../libs/jack-
son-core-
2.8.10.jar:/software/kafka_2.11-0.11.0.3/bin/../libs/jack-
son-core-
2.8.11.jar:/software/kafka_2.11-0.11.0.3/bin/../libs/jack-
son-databind-
2.8.11.2.jar:/software/kafka_2.11-0.11.0.3/bin/../libs/
jackson-jaxrs-base-
2.8.11.jar:/software/kafka_2.11-0.11.0.3/bin/../libs/jack-
son-jaxrs-json-
provider-2.8.11.jar:/software/kafka_2.11-0.11.0.3/bin/../
libs/jackson-module-
jaxb-annotations-2.8.11.jar:/software/kafka_2.11-
0.11.0.3/bin/../libs/javassist-3.21.0-GA.jar:/software/
kafka_2.11-
0.11.0.3/bin/../libs/javax.annotation-api-1.2.jar:/soft-
ware/kafka_2.11-
0.11.0.3/bin/../libs/javax.inject-1.jar:/software/kafka_2.11-
0.11.0.3/bin/../libs/javax.inject-2.5.0-b05.jar:/software/
kafka_2.11-
```

```
0.11.0.3/bin/../libs/javax.servlet-api-3.1.0.jar:/soft-
ware/kafka_2.11-
0.11.0.3/bin/../libs/javax.ws.rs-api-2.0.1.jar:/software/
kafka_2.11-
0.11.0.3/bin/../libs/jersey-client-2.24.jar:/software/kafka_
2.11-
0.11.0.3/bin/../libs/jersey-common-2.24.jar:/software/kafka_
2.11-
0.11.0.3/bin/../libs/jersey-container-servlet-2.24.jar:/
software/kafka_2.11-
0.11.0.3/bin/../libs/jersey-container-servlet-core-
2.24.jar:/software/kafka_2.11-0.11.0.3/bin/../libs/jersey-
guava-
2.24.jar:/software/kafka_2.11-0.11.0.3/bin/../libs/jersey-
media-jaxb-
2.24.jar:/software/kafka_2.11-0.11.0.3/bin/../libs/jersey-
server-
2.24.jar:/software/kafka_2.11-0.11.0.3/bin/../libs/jetty-
continuation-
9.2.22.v20170606.jar:/software/kafka_2.11-0.11.0.3/bin/../
libs/jetty-http-
9.2.22.v20170606.jar:/software/kafka_2.11-0.11.0.3/bin/../
libs/jetty-io-
9.2.22.v20170606.jar:/software/kafka_2.11-0.11.0.3/bin/../
libs/jetty-
security-9.2.22.v20170606.jar:/software/kafka_2.11-
0.11.0.3/bin/../libs/jetty-server-9.2.22.v20170606.jar:/
software/kafka_2.11-
0.11.0.3/bin/../libs/jetty-servlet-9.2.22.v20170606.jar:/
software/kafka_2.11-
0.11.0.3/bin/../libs/jetty-servlets-
9.2.22.v20170606.jar:/software/kafka_2.11-0.11.0.3/bin/../
libs/jetty-util-
9.2.22.v20170606.jar:/software/kafka_2.11-0.11.0.3/bin/../
libs/jopt-simple-
5.0.3.jar:/software/kafka_2.11-0.11.0.3/bin/../libs/kafka_
2.11-
```

```
0.11.0.3.jar:/software/kafka_2.11-0.11.0.3/bin/../libs/kafka_
2.11-0.11.0.3-
sources.jar:/software/kafka_2.11-0.11.0.3/bin/../libs/kafka_
2.11-0.11.0.3-
test-sources.jar:/software/kafka_2.11-0.11.0.3/bin/../
libs/kafka-clients-
0.11.0.3.jar:/software/kafka_2.11-0.11.0.3/bin/../libs/kafka-
log4j-appender-
0.11.0.3.jar:/software/kafka_2.11-0.11.0.3/bin/../libs/kafka-
streams-
0.11.0.3.jar:/software/kafka_2.11-0.11.0.3/bin/../libs/kafka-
streams-
examples-0.11.0.3.jar:/software/kafka_2.11-0.11.0.3/
bin/../libs/kafka-tools-
0.11.0.3.jar:/software/kafka_2.11-0.11.0.3/bin/../libs/
log4j-
1.2.17.jar:/software/kafka_2.11-0.11.0.3/bin/../libs/lz4-
1.3.0.jar:/software/kafka_2.11-0.11.0.3/bin/../libs/maven-
artifact-
3.5.0.jar:/software/kafka_2.11-0.11.0.3/bin/../libs/met-
rics-core-
2.2.0.jar:/software/kafka_2.11-0.11.0.3/bin/../libs/osgi-
resource-locator-
1.0.1.jar:/software/kafka_2.11-0.11.0.3/bin/../libs/plexus-
utils-
3.0.24.jar:/software/kafka_2.11-0.11.0.3/bin/../libs/re-
flections-
0.9.11.jar:/software/kafka_2.11-0.11.0.3/bin/../libs/rocks-
dbjni-
5.0.1.jar:/software/kafka_2.11-0.11.0.3/bin/../libs/scala-
library-
2.11.11.jar:/software/kafka_2.11-0.11.0.3/bin/../libs/scala-
parser-
combinators_2.11-1.0.4.jar:/software/kafka_2.11-0.11.0.3/
bin/../libs/slf4j-
api-1.7.25.jar:/software/kafka_2.11-0.11.0.3/bin/../libs/
slf4j-log4j12-
```

```
1.7.25.jar:/software/kafka_2.11-0.11.0.3/bin/../libs/snappy-java-
1.1.2.6.jar:/software/kafka_2.11-0.11.0.3/bin/../libs/validation-api-
1.1.0.Final.jar:/software/kafka_2.11-0.11.0.3/bin/../libs/zkclient-
0.10.jar:/software/kafka_2.11-0.11.0.3/bin/../libs/zookeeper-
3.4.10.jar
(org.apache.zookeeper.ZooKeeper)[2019-05-10 21:38:17,125] INFO Client environment:java.library.path=/usr/java/packages/lib/amd64:/usr/lib64:/lib64:/lib:/usr/lib (org.apache.zookeeper.ZooKeeper)
[2019-05-10 21:38:17,125] INFO Client environment:java.io.tmpdir=/tmp (org.apache.zookeeper.ZooKeeper)
[2019-05-10 21:38:17,125] INFO Client environment:java.compiler=<NA> (org.apache.zookeeper.ZooKeeper)
......
......
[2019-05-10 21:38:19,005] INFO [Group Metadata Manager on Broker 0]: Finished loading offsets and group metadata from __consumer_offsets-10 in 3 milliseconds. (kafka.coordinator.group.GroupMetadataManager)
[2019-05-10 21:38:19,005] INFO [Group Metadata Manager on Broker 0]: Finished loading offsets and group metadata from __consumer_offsets-13 in 0 milliseconds. (kafka.coordinator.group.GroupMetadataManager)
[2019-05-10 21:38:19,018] INFO [Group Metadata Manager on Broker 0]: Finished loading offsets and group metadata from __consumer_offsets-16 in 12 milliseconds. (kafka.coordinator.group.GroupMetadataManager)
[2019-05-10 21:38:19,018] INFO [Group Metadata Manager on Broker 0]: Finished loading offsets and group metadata from __consumer_offsets-19 in 0 milliseconds. (kafka.coordinator.group.GroupMetadataManager)
......
......
```

在 mynode2 上单独启动 Kafka，执行如下命令：

```
[root@mynode2 ~]# cd /software/kafka_2.11-0.11.0.3/bin/
[root@mynode2 bin]# ./kafka-server-start.sh /software/kafka_2.11-0.11.0.3/config/server.properties
[2019-05-10 21:46:51,009] INFO KafkaConfig values:
advertised.host.name = null
advertised.listeners = null
advertised.port = null
alter.config.policy.class.name = null
authorizer.class.name =
auto.create.topics.enable = true
auto.leader.rebalance.enable = true
background.threads = 10
broker.id = 1
broker.id.generation.enable = true
broker.rack = null
compression.type = producer
connections.max.idle.ms = 600000
controlled.shutdown.enable = true
controlled.shutdown.max.retries = 3
controlled.shutdown.retry.backoff.ms = 5000
controller.socket.timeout.ms = 30000
create.topic.policy.class.name = null
default.replication.factor = 1
delete.records.purgatory.purge.interval.requests = 1
delete.topic.enable = true
fetch.purgatory.purge.interval.requests = 1000
group.initial.rebalance.delay.ms = 0
group.max.session.timeout.ms = 300000
group.min.session.timeout.ms = 6000
host.name =
inter.broker.listener.name = null
inter.broker.protocol.version = 0.11.0-IV2
leader.imbalance.check.interval.seconds = 300
leader.imbalance.per.broker.percentage = 10
listener.security.protocol.map = SSL:SSL,SASL_PLAINTEXT:SASL_PLAINTEXT,TRACE:TRACE,SASL_SSL:SASL_SSL,PLAINTEXT:PLAINTEXT
listeners = null
log.cleaner.backoff.ms = 15000
```

```
log.cleaner.dedupe.buffer.size = 134217728
log.cleaner.delete.retention.ms = 86400000
log.cleaner.enable = true
log.cleaner.io.buffer.load.factor = 0.9
log.cleaner.io.buffer.size = 524288
log.cleaner.io.max.bytes.per.second = 1.7976931348623157E308
log.cleaner.min.cleanable.ratio = 0.5
log.cleaner.min.compaction.lag.ms = 0
log.cleaner.threads = 1
log.cleanup.policy = [delete]
log.dir = /tmp/kafka-logs
log.dirs = /kafka-logs
log.flush.interval.messages = 9223372036854775807
log.flush.interval.ms = null
log.flush.offset.checkpoint.interval.ms = 60000
log.flush.scheduler.interval.ms = 9223372036854775807
log.flush.start.offset.checkpoint.interval.ms = 60000
log.index.interval.bytes = 4096
log.index.size.max.bytes = 10485760
log.message.format.version = 0.11.0-IV2
log.message.timestamp.difference.max.ms = 9223372036854775807
log.message.timestamp.type = CreateTime
log.preallocate = false
log.retention.bytes = -1
log.retention.check.interval.ms = 300000
log.retention.hours = 168
log.retention.minutes = null
log.retention.ms = null
log.roll.hours = 168
log.roll.jitter.hours = 0
log.roll.jitter.ms = null
log.roll.ms = null
log.segment.bytes = 1073741824
log.segment.delete.delay.ms = 60000
max.connections.per.ip = 2147483647
max.connections.per.ip.overrides =
message.max.bytes = 1000012
metric.reporters = []
```

```
75.metrics.num.samples=2
76.metrics.recording.level=INFO
77.metrics.sample.window.ms=30000
78.min.insync.replicas=1
79.num.io.threads=8
80.num.network.threads=3
81.num.partitions=1
82.num.recovery.threads.per.data.dir=1
83.num.replica.fetchers=1
84.offset.metadata.max.bytes=4096
85.offsets.commit.required.acks=-1
86.offsets.commit.timeout.ms=5000
87.offsets.load.buffer.size=5242880
88.offsets.retention.check.interval.ms=600000
89.offsets.retention.minutes=1440
90.offsets.topic.compression.codec=0
91.offsets.topic.num.partitions=50
92.offsets.topic.replication.factor=1
93.offsets.topic.segment.bytes=104857600
94.port=9092
95.principal.builder.class=class org.apache.kafka.common.secur-
   ity.auth.DefaultPrincipalBuilder
96.producer.purgatory.purge.interval.requests=1000
97.queued.max.requests=500
98.quota.consumer.default=9223372036854775807
99.quota.producer.default=9223372036854775807
100.quota.window.num=11
101.quota.window.size.seconds=1
102.replica.fetch.backoff.ms=1000
103.replica.fetch.max.bytes=1048576
104.replica.fetch.min.bytes=1
105.replica.fetch.response.max.bytes=10485760
106.replica.fetch.wait.max.ms=500
107.replica.high.watermark.checkpoint.interval.ms=5000
108.replica.lag.time.max.ms=10000
109.replica.socket.receive.buffer.bytes=65536
110.replica.socket.timeout.ms=30000
111.replication.quota.window.num=11
```

```
replication.quota.window.size.seconds =1
request.timeout.ms =30000
reserved.broker.max.id =1000
sasl.enabled.mechanisms =[GSSAPI]
sasl.kerberos.kinit.cmd = /usr/bin/kinit
sasl.kerberos.min.time.before.relogin =60000
sasl.kerberos.principal.to.local.rules =[DEFAULT]
sasl.kerberos.service.name =null
sasl.kerberos.ticket.renew.jitter =0.05
sasl.kerberos.ticket.renew.window.factor =0.8
sasl.mechanism.inter.broker.protocol =GSSAPI
security.inter.broker.protocol =PLAINTEXT
socket.receive.buffer.bytes =102400
socket.request.max.bytes =104857600
socket.send.buffer.bytes =102400
ssl.cipher.suites =null
ssl.client.auth =none
ssl.enabled.protocols =[TLSv1.2, TLSv1.1, TLSv1]
ssl.endpoint.identification.algorithm =null
ssl.key.password =null
ssl.keymanager.algorithm =SunX509
ssl.keystore.location =null
ssl.keystore.password =null
ssl.keystore.type =JKS
ssl.protocol =TLS
ssl.provider =null
ssl.secure.random.implementation =null
ssl.trustmanager.algorithm =PKIX
ssl.truststore.location =null
ssl.truststore.password =null
ssl.truststore.type =JKS
transaction.abort.timed.out.transaction.cleanup.interval.ms =60000
transaction.max.timeout.ms =900000
transaction.remove.expired.transaction.cleanup.interval.ms =3600000
transaction.state.log.load.buffer.size =5242880
transaction.state.log.min.isr =1
```

```
transaction.state.log.num.partitions = 50
transaction.state.log.replication.factor = 1
transaction.state.log.segment.bytes = 104857600
transactional.id.expiration.ms = 604800000
unclean.leader.election.enable = false
zookeeper.connect = mynode3:2181,mynode4:2181,mynode5:2181
zookeeper.connection.timeout.ms = 6000
zookeeper.session.timeout.ms = 6000
zookeeper.set.acl = false
zookeeper.sync.time.ms = 2000
(kafka.server.KafkaConfig)
[2019-05-10 21:46:51,049] INFO starting (kafka.server.KafkaServer)
[2019-05-10 21:46:51,050] INFO Connecting to zookeeper on mynode3:2181,mynode4:2181,mynode5:2181 (kafka.server.KafkaServer)
[2019-05-10 21:46:51,069] INFO Starting ZkClient event thread. (org.I0Itec.zkclient.ZkEventThread)
[2019-05-10 21:46:51,082] INFO Client environment:zookeeper.version=3.4.10-39d3a4f269333c922ed3db283be479f9deacaa0f, built on 03/23/2017 10:13 GMT (org.apache.zookeeper.ZooKeeper)
[2019-05-10 21:46:51,082] INFO Client environment:host.name=mynode2 (org.apache.zookeeper.ZooKeeper)
[2019-05-10 21:46:51,082] INFO Client environment:java.version=1.8.0_181 (org.apache.zookeeper.ZooKeeper)
[2019-05-10 21:46:51,082] INFO Client environment:java.vendor=Oracle Corporation (org.apache.zookeeper.ZooKeeper)
[2019-05-10 21:46:51,082] INFO Client environment:java.home=/software/jdk1.8.0_181/jre (org.apache.zookeeper.ZooKeeper)
[2019-05-10 21:46:51,082] INFO Client environment:java.class.path=:/software/kafka_2.11-0.11.0.3/bin/../libs/aopalliance-repackaged-2.5.0-b05.jar:/software/kafka_2.11-0.11.0.3/bin/../libs/a
rgparse4j-0.7.0.jar:/software/kafka_2.11-0.11.0.3/bin/../libs/commons-lang3-
3.5.jar:/software/kafka_2.11-0.11.0.3/bin/../libs/connect-api-
```

```
0.11.0.3.jar:/software/kafka_2.11-0.11.0.3/bin/../libs/con-
nect-file-
0.11.0.3.jar:/software/kafka_2.11-0.11.0.3/bin/../libs/con-
nect-json-
0.11.0.3.jar:/software/kafka_2.11-0.11.0.3/bin/../libs/con-
nect-runtime-
0.11.0.3.jar:/software/kafka_2.11-0.11.0.3/bin/../libs/con-
nect-transforms-
0.11.0.3.jar:/software/kafka_2.11-0.11.0.3/bin/../libs/guava-
20.0.jar:/software/kafka_2.11-0.11.0.3/bin/../libs/hk2-
api-2.5.0-
b05.jar:/software/kafka_2.11-0.11.0.3/bin/../libs/hk2-lo-
cator-2.5.0-
b05.jar:/software/kafka_2.11-0.11.0.3/bin/../libs/hk2-
utils-2.5.0-
b05.jar:/software/kafka_2.11-0.11.0.3/bin/../libs/jackson-
annotations-
2.8.11.jar:/software/kafka_2.11-0.11.0.3/bin/../libs/jack-
son-core-
2.8.10.jar:/software/kafka_2.11-0.11.0.3/bin/../libs/jack-
son-core-
2.8.11.jar:/software/kafka_2.11-0.11.0.3/bin/../libs/jack-
son-databind-
2.8.11.2.jar:/software/kafka_2.11-0.11.0.3/bin/../libs/
jackson-jaxrs-base-
2.8.11.jar:/software/kafka_2.11-0.11.0.3/bin/../libs/jack-
son-jaxrs-json-
provider-2.8.11.jar:/software/kafka_2.11-0.11.0.3/bin/../
libs/jackson-module-
jaxb-annotations-2.8.11.jar:/software/kafka_2.11-
0.11.0.3/bin/../libs/javassist-3.21.0-GA.jar:/software/
kafka_2.11-
0.11.0.3/bin/../libs/javax.annotation-api-1.2.jar:/soft-
ware/kafka_2.11-
0.11.0.3/bin/../libs/javax.inject-1.jar:/software/kafka_2.11-
0.11.0.3/bin/../libs/javax.inject-2.5.0-b05.jar:/software/
kafka_2.11-
```

```
0.11.0.3/bin/../libs/javax.servlet-api-3.1.0.jar:/soft-
ware/kafka_2.11-
0.11.0.3/bin/../libs/javax.ws.rs-api-2.0.1.jar:/software/
kafka_2.11-
0.11.0.3/bin/../libs/jersey-client-2.24.jar:/software/kafka_
2.11-
0.11.0.3/bin/../libs/jersey-common-2.24.jar:/software/kafka_
2.11-
0.11.0.3/bin/../libs/jersey-container-servlet-2.24.jar:/
software/kafka_2.11-
0.11.0.3/bin/../libs/jersey-container-servlet-core-
2.24.jar:/software/kafka_2.11-0.11.0.3/bin/../libs/jersey-
guava-
2.24.jar:/software/kafka_2.11-0.11.0.3/bin/../libs/jersey-
media-jaxb-
2.24.jar:/software/kafka_2.11-0.11.0.3/bin/../libs/jersey-
server-
2.24.jar:/software/kafka_2.11-0.11.0.3/bin/../libs/jetty-
continuation-
9.2.22.v20170606.jar:/software/kafka_2.11-0.11.0.3/bin/../
libs/jetty-http-
9.2.22.v20170606.jar:/software/kafka_2.11-0.11.0.3/bin/../
libs/jetty-io-
9.2.22.v20170606.jar:/software/kafka_2.11-0.11.0.3/bin/../
libs/jetty-
security-9.2.22.v20170606.jar:/software/kafka_2.11-
0.11.0.3/bin/../libs/jetty-server-9.2.22.v20170606.jar:/
software/kafka_2.11-
0.11.0.3/bin/../libs/jetty-servlet-9.2.22.v20170606.jar:/
software/kafka_2.11-
0.11.0.3/bin/../libs/jetty-servlets-
9.2.22.v20170606.jar:/software/kafka_2.11-0.11.0.3/bin/../
libs/jetty-util-
9.2.22.v20170606.jar:/software/kafka_2.11-0.11.0.3/bin/../
libs/jopt-simple-
5.0.3.jar:/software/kafka_2.11-0.11.0.3/bin/../libs/kafka_
2.11-
```

```
0.11.0.3.jar:/software/kafka_2.11-0.11.0.3/bin/../libs/kafka_
2.11-0.11.0.3-
sources.jar:/software/kafka_2.11-0.11.0.3/bin/../libs/kafka_
2.11-0.11.0.3-
test-sources.jar:/software/kafka_2.11-0.11.0.3/bin/../
libs/kafka-clients-
0.11.0.3.jar:/software/kafka_2.11-0.11.0.3/bin/../libs/kafka-
log4j-appender-
0.11.0.3.jar:/software/kafka_2.11-0.11.0.3/bin/../libs/kafka-
streams-
0.11.0.3.jar:/software/kafka_2.11-0.11.0.3/bin/../libs/kafka-
streams-
examples-0.11.0.3.jar:/software/kafka_2.11-0.11.0.3/
bin/../libs/kafka-tools-
0.11.0.3.jar:/software/kafka_2.11-0.11.0.3/bin/../libs/
log4j-
1.2.17.jar:/software/kafka_2.11-0.11.0.3/bin/../libs/lz4-
1.3.0.jar:/software/kafka_2.11-0.11.0.3/bin/../libs/maven-
artifact-
3.5.0.jar:/software/kafka_2.11-0.11.0.3/bin/../libs/met-
rics-core-
2.2.0.jar:/software/kafka_2.11-0.11.0.3/bin/../libs/osgi-
resource-locator-
1.0.1.jar:/software/kafka_2.11-0.11.0.3/bin/../libs/plexus-
utils-
3.0.24.jar:/software/kafka_2.11-0.11.0.3/bin/../libs/re-
flections-
0.9.11.jar:/software/kafka_2.11-0.11.0.3/bin/../libs/rocks-
dbjni-
5.0.1.jar:/software/kafka_2.11-0.11.0.3/bin/../libs/scala-
library-
2.11.11.jar:/software/kafka_2.11-0.11.0.3/bin/../libs/scala-
parser-
combinators_2.11-1.0.4.jar:/software/kafka_2.11-0.11.0.3/
bin/../libs/slf4j-
api-1.7.25.jar:/software/kafka_2.11-0.11.0.3/bin/../libs/
slf4j-log4j12-
```

```
    1.7.25.jar:/software/kafka_2.11-0.11.0.3/bin/../libs/snappy-
    java-
    1.1.2.6.jar:/software/kafka_2.11-0.11.0.3/bin/../libs/vali-
    dation-api-
    1.1.0.Final.jar:/software/kafka_2.11-0.11.0.3/bin/../libs/
    zkclient-
    0.10.jar:/software/kafka_2.11-0.11.0.3/bin/../libs/zookeeper-
    3.4.10.jar
    (org.apache.zookeeper.ZooKeeper)[2019-05-10 21:46:51,082]
    INFO Client environment:java.library.path=/usr/java/packa-
    ges/lib/amd64:/usr/lib64:/lib64:/lib:/usr/lib (org.apache.
    zookeeper.ZooKeeper)
169.[2019-05-10 21:46:51,082] INFO Client environment:java.io.
    tmpdir=/tmp (org.apache.zookeeper.ZooKeeper)
170.[2019-05-10 21:46:51,082] INFO Client environment:java.com-
    piler=<NA> (org.apache.zookeeper.ZooKeeper)
171.[2019-05-10 21:46:51,082] INFO Client environment:os.name=
    Linux (org.apache.zookeeper.ZooKeeper)
172.......
173.......
174.[2019-05-10 21:46:52,619] INFO Partition [__consumer_off-
    sets,11] on broker 1: __consumer_offsets-11 starts at Leader
    Epoch 31 from offset 0. Previous Leader Epoch was: 30 (kafka.
    cluster.Partition)
175.[2019-05-10 21:46:52,619] INFO Partition [__consumer_off-
    sets,8] on broker 1: __consumer_offsets-8 starts at Leader Ep-
    och 31 from offset 97. Previous Leader Epoch was: 30 (kafka.
    cluster.Partition)
176.[2019-05-10 21:46:52,619] INFO Partition [__consumer_off-
    sets,5] on broker 1: __consumer_offsets-5 starts at Leader Ep-
    och 31 from offset 0. Previous Leader Epoch was: 30 (kafka.clus-
    ter.Partition)
177.[2019-05-10 21:46:52,619] INFO Partition [__consumer_off-
    sets,2] on broker 1: __consumer_offsets-2 starts at Leader Ep-
    och 31 from offset 0. Previous Leader Epoch was: 30 (kafka.clus-
    ter.Partition)
178.[2019-05-10 21:46:52,619] INFO Partition [__consumer_offsets,
```

```
    47] on broker 1: __consumer_offsets -47 starts at Leader Epoch
    31 from offset 714. Previous Leader Epoch was: 30 (kafka.clus-
    ter.Partition)
179.[2019-05-10 21:46:52,619] INFO Partition [__consumer_off-
    sets,38] on broker 1: __consumer_offsets -38 starts at Leader
    Epoch 31 from offset 0. Previous Leader Epoch was: 30 (kafka.
    cluster.Partition)
180.[2019-05-10 21:46:52,620] INFO Partition [RoadRealTimeLog,
    0] on broker 1: RoadRealTimeLog-0 starts at Leader Epoch 2 from
    offset 3102. Previous Leader Epoch was: 1 (kafka.cluster.Parti-
    tion)
181.[2019-05-10 21:46:52,620] INFO Partition [__consumer_off-
    sets,35] on broker 1: __consumer_offsets -35 starts at Leader
    Epoch 31 from offset 0. Previous Leader Epoch was: 30 (kafka.
    cluster.Partition)
182.[2019-05-10 21:46:52,620] INFO Partition [__consumer_off-
    sets,44] on broker 1: __consumer_offsets -44 starts at Leader
    Epoch 31 from offset 0. Previous Leader Epoch was: 30 (kafka.
    cluster.Partition)
183.[2019-05-10 21:46:52,620] INFO Partition [__consumer_off-
    sets,32] on broker 1: __consumer_offsets -32 starts at Leader
    Epoch 31 from offset 0. Previous Leader Epoch was: 30 (kafka.
    cluster.Partition)
184.[2019-05-10 21:46:52,620] INFO Partition [__consumer_off-
    sets,41] on broker 1: __consumer_offsets -41 starts at Leader
    Epoch 31 from offset 0. Previous Leader Epoch was: 30 (kafka.
    cluster.Partition)
185.[2019-05-10 21:46:52,630] INFO [Group Metadata Manager on
    Broker 1]: Finished loading offsets and group metadata from __
    consumer_offsets-8 in 27 milliseconds. (kafka.coordinator.
    group.GroupMetadataManager)
186.[2019-05-10 21:46:52,630] INFO [Group Metadata Manager on
    Broker 1]: Finished loading offsets and group metadata from __
    consumer_offsets -11 in 0 milliseconds. (kafka.coordinator.
    group.Group MetadataManager)
187.[2019-05-10 21:46:52,630] INFO [Group Metadata Manager on
    Broker 1]: Finished loading offsets and group
```

```
     metadata from __consumer_offsets-14 in 0 milliseconds. (kafka.
     coordinator.group.Group MetadataManager)
188. [2019-05-10 21:46:52,635] INFO [Group Metadata Manager on
     Broker 1]: Finished loading offsets and group metadata from __
     consumer_offsets-17 in 5 milliseconds. (kafka.coordinator.
     group.GroupMetadataManager)
189. [2019-05-10 21:46:52,635] INFO [Group Metadata Manager on
     Broker 1]: Finished loading offsets and group metadata from __
     consumer_offsets-20 in 0 milliseconds. (kafka.coordinator.
     group.GroupMetadataManager)
190. [2019-05-10 21:46:52,636] INFO [Group Metadata Manager on
     Broker 1]: Finished loading offsets and group metadata from __
     consumer_offsets-23 in 0 milliseconds. (kafka.coordinator.
     group.GroupMetadataManager)
191. [2019-05-10 21:46:52,637] INFO [Group Metadata Manager on
     Broker 1]: Finished loading offsets and group metadata from __
     consumer_offsets-26 in 1 milliseconds. (kafka.coordinator.
     group.GroupMetadataManager)
192. [2019-05-10 21:46:52,654] INFO [Group Metadata Manager on
     Broker 1]: Finished loading offsets and group metadata from __
     consumer_offsets-29 in 16 milliseconds. (kafka.coordinator.
     group.GroupMetadataManager)
193. [2019-05-10 21:46:52,654] INFO [Group Metadata Manager on
     Broker 1]: Finished loading offsets and group metadata from __
     consumer_offsets-32 in 0 milliseconds. (kafka.coordinator.
     group.GroupMetadataManager)
194. [2019-05-10 21:46:52,655] INFO [Group Metadata Manager on
     Broker 1]: Finished loading offsets and group metadata from __
     consumer_offsets-35 in 0 milliseconds. (kafka.coordinator.
     group.GroupMetadataManager)
195. [2019-05-10 21:46:52,655] INFO [Group Metadata Manager on
     Broker 1]: Finished loading offsets and group metadata from __
     consumer_offsets-38 in 0 milliseconds. (kafka.coordinator.
     group.GroupMetadataManager)
196. [2019-05-10 21:46:52,656] INFO [Group Metadata Manager on
     Broker 1]: Finished loading offsets and group metadata from __
     consumer_offsets-41 in 1 milliseconds. (kafka.coordinator.
     group.GroupMetadataManager)
```

```
[2019-05-10 21:46:52,656] INFO [Group Metadata Manager on Broker 1]: Finished loading offsets and group metadata from __consumer_offsets-44 in 0 milliseconds. (kafka.coordinator.group.GroupMetadataManager)
[2019-05-10 21:46:52,670] INFO [Group Metadata Manager on Broker 1]: Finished loading offsets and group metadata from __consumer_offsets-47 in 14 milliseconds. (kafka.coordinator.group.GroupMetadataManager)
[2019-05-10 21:46:53,511] INFO Partition [flumetopic,1] on broker 1: Expanding ISR from 1 to 1,0 (kafka.cluster.Partition)
[2019-05-10 21:46:53,519] INFO Partition [firsttopic,2] on broker 1: Expanding ISR from 1 to 1,0 (kafka.cluster.Partition)
[2019-05-10 21:46:53,524] INFO Partition [mytest,2] on broker 1: Expanding ISR from 1 to 1,0 (kafka.cluster.Partition)
... ...
```

在 mynode3 上单独启动 Kafka，执行如下命令：

```
[root@mynode3 ~]# cd /software/kafka_2.11-0.11.0.3/bin/
[root@mynode3 bin]# ./kafka-server-start.sh /software/kafka_2.11-0.11.0.3/config/server.properties
[2019-05-10 21:49:04,218] INFO KafkaConfig values:
advertised.host.name = null
advertised.listeners = null
advertised.port = null
alter.config.policy.class.name = null
authorizer.class.name =
auto.create.topics.enable = true
auto.leader.rebalance.enable = true
background.threads = 10
broker.id = 2
broker.id.generation.enable = true
broker.rack = null
compression.type = producer
connections.max.idle.ms = 600000
controlled.shutdown.enable = true
controlled.shutdown.max.retries = 3
controlled.shutdown.retry.backoff.ms = 5000
```

```
controller.socket.timeout.ms=30000
create.topic.policy.class.name=null
default.replication.factor=1
delete.records.purgatory.purge.interval.requests=1
delete.topic.enable=true
fetch.purgatory.purge.interval.requests=1000
group.initial.rebalance.delay.ms=0
group.max.session.timeout.ms=300000
group.min.session.timeout.ms=6000
host.name=
inter.broker.listener.name=null
inter.broker.protocol.version=0.11.0-IV2
leader.imbalance.check.interval.seconds=300
leader.imbalance.per.broker.percentage=10
listener.security.protocol.map=SSL:SSL,SASL_PLAINTEXT:SASL_PLAINTEXT,TRACE:TRACE,SASL_SSL:SASL_SSL,PLAINTEXT:PLAINTEXT
listeners=null
log.cleaner.backoff.ms=15000
log.cleaner.dedupe.buffer.size=134217728
log.cleaner.delete.retention.ms=86400000
log.cleaner.enable=true
log.cleaner.io.buffer.load.factor=0.9
log.cleaner.io.buffer.size=524288
log.cleaner.io.max.bytes.per.second=1.7976931348623157E308
log.cleaner.min.cleanable.ratio=0.5
log.cleaner.min.compaction.lag.ms=0
log.cleaner.threads=1
log.cleanup.policy=[delete]
log.dir=/tmp/kafka-logs
log.dirs=/kafka-logs
log.flush.interval.messages=9223372036854775807
log.flush.interval.ms=null
log.flush.offset.checkpoint.interval.ms=60000
log.flush.scheduler.interval.ms=9223372036854775807
log.flush.start.offset.checkpoint.interval.ms=60000
log.index.interval.bytes=4096
log.index.size.max.bytes=10485760
log.message.format.version=0.11.0-IV2
```

```
log.message.timestamp.difference.max.ms = 9223372036854775807
log.message.timestamp.type = CreateTime
log.preallocate = false
log.retention.bytes = -1
log.retention.check.interval.ms = 300000
log.retention.hours = 168
log.retention.minutes = null
log.retention.ms = null
log.roll.hours = 168
log.roll.jitter.hours = 0
log.roll.jitter.ms = null
log.roll.ms = null
log.segment.bytes = 1073741824
log.segment.delete.delay.ms = 60000
max.connections.per.ip = 2147483647
max.connections.per.ip.overrides =
message.max.bytes = 1000012
metric.reporters = []
metrics.num.samples = 2
metrics.recording.level = INFO
metrics.sample.window.ms = 30000
min.insync.replicas = 1
num.io.threads = 8
num.network.threads = 3
num.partitions = 1
num.recovery.threads.per.data.dir = 1
num.replica.fetchers = 1
offset.metadata.max.bytes = 4096
offsets.commit.required.acks = -1
offsets.commit.timeout.ms = 5000
offsets.load.buffer.size = 5242880
offsets.retention.check.interval.ms = 600000
offsets.retention.minutes = 1440
offsets.topic.compression.codec = 0
offsets.topic.num.partitions = 50
offsets.topic.replication.factor = 1
offsets.topic.segment.bytes = 104857600
port = 9092
```

```
95. principal.builder.class = class org.apache.kafka.common.secur-
    ity.auth.DefaultPrincipalBuilder
96. producer.purgatory.purge.interval.requests = 1000
97. queued.max.requests = 500
98. quota.consumer.default = 9223372036854775807
99. quota.producer.default = 9223372036854775807
100. quota.window.num = 11
101. quota.window.size.seconds = 1
102. replica.fetch.backoff.ms = 1000
103. replica.fetch.max.bytes = 1048576
104. replica.fetch.min.bytes = 1
105. replica.fetch.response.max.bytes = 10485760
106. replica.fetch.wait.max.ms = 500
107. replica.high.watermark.checkpoint.interval.ms = 5000
108. replica.lag.time.max.ms = 10000
109. replica.socket.receive.buffer.bytes = 65536
110. replica.socket.timeout.ms = 30000
111. replication.quota.window.num = 11
112. replication.quota.window.size.seconds = 1
113. request.timeout.ms = 30000
114. reserved.broker.max.id = 1000
115. sasl.enabled.mechanisms = [GSSAPI]
116. sasl.kerberos.kinit.cmd = /usr/bin/kinit
117. sasl.kerberos.min.time.before.relogin = 60000
118. sasl.kerberos.principal.to.local.rules = [DEFAULT]
119. sasl.kerberos.service.name = null
120. sasl.kerberos.ticket.renew.jitter = 0.05
121. sasl.kerberos.ticket.renew.window.factor = 0.8
122. sasl.mechanism.inter.broker.protocol = GSSAPI
123. security.inter.broker.protocol = PLAINTEXT
124. socket.receive.buffer.bytes = 102400
125. socket.request.max.bytes = 104857600
126. socket.send.buffer.bytes = 102400
127. ssl.cipher.suites = null
128. ssl.client.auth = none
129. ssl.enabled.protocols = [TLSv1.2, TLSv1.1, TLSv1]
130. ssl.endpoint.identification.algorithm = null
131. ssl.key.password = null
```

```
ssl.keymanager.algorithm = SunX509
ssl.keystore.location = null
ssl.keystore.password = null
ssl.keystore.type = JKS
ssl.protocol = TLS
ssl.provider = null
ssl.secure.random.implementation = null
ssl.trustmanager.algorithm = PKIX
ssl.truststore.location = null
ssl.truststore.password = null
ssl.truststore.type = JKS
transaction.abort.timed.out.transaction.cleanup.interval.ms = 60000
transaction.max.timeout.ms = 900000
transaction.remove.expired.transaction.cleanup.interval.ms = 3600000
transaction.state.log.load.buffer.size = 5242880
transaction.state.log.min.isr = 1
transaction.state.log.num.partitions = 50
transaction.state.log.replication.factor = 1
transaction.state.log.segment.bytes = 104857600
transactional.id.expiration.ms = 604800000
unclean.leader.election.enable = false
zookeeper.connect = mynode3:2181,mynode4:2181,mynode5:2181
zookeeper.connection.timeout.ms = 6000
zookeeper.session.timeout.ms = 6000
zookeeper.set.acl = false
zookeeper.sync.time.ms = 2000
(kafka.server.KafkaConfig)
[2019-05-10 21:49:04,258] INFO starting (kafka.server.KafkaServer)
[2019-05-10 21:49:04,260] INFO Connecting to zookeeper on mynode3:2181,mynode4:2181,mynode5:2181 (kafka.server.KafkaServer)
[2019-05-10 21:49:04,279] INFO Starting ZkClient event thread. (org.I0Itec.zkclient.ZkEventThread)
[2019-05-10 21:49:04,284] INFO Client environment:zookeeper.
```

```
    version = 3.4.10 - 39d3a4f269333c922ed3db283be479f9deacaa0f,
    built on 03/23/2017 10:13 GMT (org.apache.zookeeper.ZooKeeper)
163. [2019 - 05 - 10 21:49:04,284] INFO Client environment:host.name
    = mynode3 (org.apache.zookeeper.ZooKeeper)
164. [2019 - 05 - 10 21:49:04,284] INFO Client environment:java.ver-
    sion = 1.8.0_181 (org.apache.zookeeper.ZooKeeper)
165. [2019 - 05 - 10 21:49:04,284] INFO Client environment:java.ven-
    dor = Oracle Corporation (org.apache.zookeeper.ZooKeeper)
166. [2019 - 05 - 10 21:49:04,284] INFO Client environment:java.home
    = /software/jdk1.8.0_181/jre (org.apache.zookeeper.ZooKeeper)
167. [2019 - 05 - 10 21:49:04,284] INFO Client environment:java.
    class.path = :/software/kafka_2.11 - 0.11.0.3/bin/../libs/aop-
    alliance - repackaged - 2.5.0 - b05.jar:/software/kafka_2.11 -
    0.11.0.3/bin/../libs/a
168. rgparse4j - 0.7.0.jar:/software/kafka_2.11 - 0.11.0.3/bin/../
    libs/commons - lang3 -
    3.5.jar:/software/kafka_2.11 - 0.11.0.3/bin/../libs/connect -
    api -
    0.11.0.3.jar:/software/kafka_2.11 - 0.11.0.3/bin/../libs/con-
    nect - file -
    0.11.0.3.jar:/software/kafka_2.11 - 0.11.0.3/bin/../libs/con-
    nect - json -
    0.11.0.3.jar:/software/kafka_2.11 - 0.11.0.3/bin/../libs/con-
    nect - runtime -
    0.11.0.3.jar:/software/kafka_2.11 - 0.11.0.3/bin/../libs/con-
    nect - transforms -
    0.11.0.3.jar:/software/kafka_2.11 - 0.11.0.3/bin/../libs/guava -
    20.0.jar:/software/kafka_2.11 - 0.11.0.3/bin/../libs/hk2 -
    api - 2.5.0 -
    b05.jar:/software/kafka_2.11 - 0.11.0.3/bin/../libs/hk2 -
    locator - 2.5.0 -
    b05.jar:/software/kafka_2.11 - 0.11.0.3/bin/../libs/hk2 -
    utils - 2.5.0 -
    b05.jar:/software/kafka_2.11 - 0.11.0.3/bin/../libs/jackson -
    annotations -
    2.8.11.jar:/software/kafka_2.11 - 0.11.0.3/bin/../libs/jack-
    son - core -
```

```
2.8.10.jar:/software/kafka_2.11-0.11.0.3/bin/../libs/jack-
son-core-
2.8.11.jar:/software/kafka_2.11-0.11.0.3/bin/../libs/jack-
son-databind-
2.8.11.2.jar:/software/kafka_2.11-0.11.0.3/bin/../libs/
jackson-jaxrs-base-
2.8.11.jar:/software/kafka_2.11-0.11.0.3/bin/../libs/jack-
son-jaxrs-json-
provider-2.8.11.jar:/software/kafka_2.11-0.11.0.3/bin/../
libs/jackson-module-
jaxb-annotations-2.8.11.jar:/software/kafka_2.11-
0.11.0.3/bin/../libs/javassist-3.21.0-GA.jar:/software/
kafka_2.11-
0.11.0.3/bin/../libs/javax.annotation-api-1.2.jar:/soft-
ware/kafka_2.11-
0.11.0.3/bin/../libs/javax.inject-1.jar:/software/kafka_2.11-
0.11.0.3/bin/../libs/javax.inject-2.5.0-b05.jar:/software/
kafka_2.11-
0.11.0.3/bin/../libs/javax.servlet-api-3.1.0.jar:/soft-
ware/kafka_2.11-
0.11.0.3/bin/../libs/javax.ws.rs-api-2.0.1.jar:/software/
kafka_2.11-
0.11.0.3/bin/../libs/jersey-client-2.24.jar:/software/kafka_
2.11-
0.11.0.3/bin/../libs/jersey-common-2.24.jar:/software/kafka_
2.11-
0.11.0.3/bin/../libs/jersey-container-servlet-2.24.jar:/
software/kafka_2.11-
0.11.0.3/bin/../libs/jersey-container-servlet-core-
2.24.jar:/software/kafka_2.11-0.11.0.3/bin/../libs/jersey-
guava-
2.24.jar:/software/kafka_2.11-0.11.0.3/bin/../libs/jersey-
media-jaxb-
2.24.jar:/software/kafka_2.11-0.11.0.3/bin/../libs/jersey-
server-
2.24.jar:/software/kafka_2.11-0.11.0.3/bin/../libs/jetty-
continuation-
```

```
9.2.22.v20170606.jar:/software/kafka_2.11-0.11.0.3/bin/../
libs/jetty-http-
9.2.22.v20170606.jar:/software/kafka_2.11-0.11.0.3/bin/../
libs/jetty-io-
9.2.22.v20170606.jar:/software/kafka_2.11-0.11.0.3/bin/../
libs/jetty-security-
9.2.22.v20170606.jar:/software/kafka_2.11-0.11.0.3/bin/../
libs/jetty-server-
9.2.22.v20170606.jar:/software/kafka_2.11-0.11.0.3/bin/../
libs/jetty-servlet-
9.2.22.v20170606.jar:/software/kafka_2.11-0.11.0.3/bin/../
libs/jetty-servlets-
9.2.22.v20170606.jar:/software/kafka_2.11-0.11.0.3/bin/../
libs/jetty-util-
9.2.22.v20170606.jar:/software/kafka_2.11-0.11.0.3/bin/../
libs/jopt-simple-
5.0.3.jar:/software/kafka_2.11-0.11.0.3/bin/../libs/kafka_
2.11-
0.11.0.3.jar:/software/kafka_2.11-0.11.0.3/bin/../libs/kafka_
2.11-0.11.0.3-
sources.jar:/software/kafka_2.11-0.11.0.3/bin/../libs/kafka_
2.11-0.11.0.3-
test-sources.jar:/software/kafka_2.11-0.11.0.3/bin/../
libs/kafka-clients-
0.11.0.3.jar:/software/kafka_2.11-0.11.0.3/bin/../libs/kafka-
log4j-appender-
0.11.0.3.jar:/software/kafka_2.11-0.11.0.3/bin/../libs/kafka-
streams-
0.11.0.3.jar:/software/kafka_2.11-0.11.0.3/bin/../libs/kaf-
ka-streams-examples-
0.11.0.3.jar:/software/kafka_2.11-0.11.0.3/bin/../libs/kafka-
tools-
0.11.0.3.jar:/software/kafka_2.11-0.11.0.3/bin/../libs/
log4j-
1.2.17.jar:/software/kafka_2.11-0.11.0.3/bin/../libs/lz4-
1.3.0.jar:/software/kafka_2.11-0.11.0.3/bin/../libs/maven-
artifact-
```

```
    3.5.0.jar:/software/kafka_2.11-0.11.0.3/bin/../libs/met-
    rics-core-
    2.2.0.jar:/software/kafka_2.11-0.11.0.3/bin/../libs/osgi-
    resource-locator-
    1.0.1.jar:/software/kafka_2.11-0.11.0.3/bin/../libs/plexus-
    utils-
    3.0.24.jar:/software/kafka_2.11-0.11.0.3/bin/../libs/
    reflections-
    0.9.11.jar:/software/kafka_2.11-0.11.0.3/bin/../libs/rocks-
    dbjni-
    5.0.1.jar:/software/kafka_2.11-0.11.0.3/bin/../libs/scala-
    library-
    2.11.11.jar:/software/kafka_2.11-0.11.0.3/bin/../libs/scala-
    parser-
    combinators_2.11-1.0.4.jar:/software/kafka_2.11-0.11.0.3/
    bin/../libs/slf4j-
    api-1.7.25.jar:/software/kafka_2.11-0.11.0.3/bin/../libs/
    slf4j-log4j12-
    1.7.25.jar:/software/kafka_2.11-0.11.0.3/bin/../libs/snappy-
    java-
    1.1.2.6.jar:/software/kafka_2.11-0.11.0.3/bin/../libs/vali-
    dation-api-
    1.1.0.Final.jar:/software/kafka_2.11-0.11.0.3/bin/../libs/
    zkclient-
    0.10.jar:/software/kafka_2.11-0.11.0.3/bin/../libs/zookeeper-
    3.4.10.jar
    (org.apache.zookeeper.ZooKeeper)[2019-05-10 21:49:04,285]
    INFO Client environment:java.library.path=/usr/java/packa-
    ges/lib/amd64:/usr/lib64:/lib64:/lib:/usr/lib (org.apache.
    zookeeper.ZooKeeper)
169. [2019-05-10 21:49:04,285] INFO Client environment:java.io.
    tmpdir=/tmp (org.apache.zookeeper.ZooKeeper)
170. [2019-05-10 21:49:04,285] INFO Client environment:java.com-
    piler=<NA> (org.apache.zookeeper.ZooKeeper)
171. [2019-05-10 21:49:04,285] INFO Client environment:os.name=
    Linux (org.apache.zookeeper.ZooKeeper)
172. [2019-05-10 21:49:04,285] INFO Client environment:os.arch=
    amd64 (org.apache.zookeeper.ZooKeeper)
```

```
[2019-05-10 21:49:04,285] INFO Client environment:os.version=2.6.32-504.el6.x86_64 (org.apache.zookeeper.ZooKeeper)
[2019-05-10 21:49:04,285] INFO Client environment:user.name=root (org.apache.zookeeper.ZooKeeper)
[2019-05-10 21:49:04,285] INFO Client environment:user.home=/root (org.apache.zookeeper.ZooKeeper)
[2019-05-10 21:49:04,285] INFO Client environment:user.dir=/software/kafka_2.11-0.11.0.3/bin (org.apache.zookeeper.ZooKeeper)
[2019-05-10 21:49:04,286] INFO Initiating client connection, connectString=mynode3:2181,mynode4:2181,mynode5:2181 sessionTimeout=6000 watcher=org.I0Itec.zkclient.ZkClient@1fa268de (org.apache.zookeeper.ZooKeeper)
[2019-05-10 21:49:04,301] INFO Waiting for keeper state SyncConnected (org.I0Itec.zkclient.ZkClient)
[2019-05-10 21:49:04,303] INFO Opening socket connection to server mynode4/192.168.179.16:2181. Will not attempt to authenticate using SASL (unknown error) (org.apache.zookeeper.ClientCnxn)
[2019-05-10 21:49:04,309] INFO Socket connection established to mynode4/192.168.179.16:2181, initiating session (org.apache.zookeeper.ClientCnxn)
[2019-05-10 21:49:04,328] INFO Session establishment complete on server mynode4/192.168.179.16:2181, sessionid=0x26aa1f2ef190001, negotiated timeout=6000 (org.apache.zookeeper.ClientCnxn)
[2019-05-10 21:49:04,330] INFO zookeeper state changed (SyncConnected) (org.I0Itec.zkclient.ZkClient)
......
......
[2019-05-10 21:49:05,970] INFO [Group Metadata Manager on Broker 2]: Finished loading offsets and group metadata from __consumer_offsets-45 in 8 milliseconds. (kafka.coordinator.group.GroupMetadataManager)
[2019-05-10 21:49:05,970] INFO [Group Metadata Manager on Broker 2]: Finished loading offsets and group metadata from __
```

```
consumer_offsets - 48 in 0 milliseconds. (kafka.coordinator.group.Group
MetadataManager)[2019 -05 -10 21:49:05,973] INFO [Group Metadata Manager on Broker 2]: Finished loading offsets and group metadata from __consumer_offsets -3 in 3 milliseconds. (kafka.coordinator.group.GroupM
etadataManager) [2019 - 05 - 10 21:49:06,387] INFO Partition [mytest,1] on broker 2: Expanding ISR from 2 to 2,0 (kafka.cluster.Partition)
[2019 -05 -10 21:49:06,403] INFO Partition [mytest,0] on broker 2: Expanding ISR from 2 to 2,0 (kafka.cluster.Partition)
[2019 -05 -10 21:49:06,409] INFO Partition [firsttopic,1] on broker 2: Expanding ISR from 2 to 2,0 (kafka.cluster.Partition)
[2019 -05 -10 21:49:06,413] INFO Partition [firsttopic,0] on broker 2: Expanding ISR from 2 to 2,0 (kafka.cluster.Partition)
[2019 -05 -10 21:49:06,419] INFO Partition [flumetopic,2] on broker 2: Expanding ISR from 2 to 2,0 (kafka.cluster.Partition)
[2019 -05 -10 21:49:06,424] INFO Partition [mytest,1] on broker 2: Expanding ISR from 2,0 to 2,0,1 (kafka.cluster.Partition)
[2019 -05 -10 21:49:06,428] INFO Partition [flumetopic,0] on broker 2: Expanding ISR from 2 to 2,0 (kafka.cluster.Partition)
[2019 -05 -10 21:49:06,432] INFO Partition [mytest,0] on broker 2: Expanding ISR from 2,0 to 2,0,1 (kafka.cluster.Partition)
[2019 -05 -10 21:49:06,437] INFO Partition [firsttopic,1] on broker 2: Expanding ISR from 2,0 to 2,0,1 (kafka.cluster.Partition)
[2019 -05 -10 21:49:06,440] INFO Partition [firsttopic,0] on broker 2: Expanding ISR from 2,0 to 2,0,1 (kafka.cluster.Partition)
[2019 -05 -10 21:49:06,444] INFO Partition [flumetopic,2] on broker 2: Expanding ISR from 2,0 to 2,0,1 (kafka.cluster.Partition)
[2019 -05 -10 21:49:06,447] INFO Partition [flumetopic,0] on broker 2: Expanding ISR from 2,0 to 2,0,1 (kafka.cluster.Partition)
... ...
```

经过以上步骤的配置及重启 Kafka 集群，可以删除 Kafka 中的“firsttopic”消息队列。下面以在 mynode1 中执行删除命令为例，执行如下命令：

```
[root@mynode1 ~]# cd /software/kafka_2.11 -0.11.0.3/bin/
[root@mynode1 bin]# ./kafka - topics.sh --zookeeper mynode3:2181,mynode4:2181,mynode5:2181 --delete --topic firsttopic
```

```
3. Topic firsttopic is marked for deletion.
4. Note: This will have no impact if delete.topic.enable is not set
   to true.
```

经过以上删除“firsttopic”消息队列命令的执行，可以在任意一台 Kafka Broker 节点上执行如下命令，验证是否成功将“firsttopic”删除。

```
1. [root@mynode1 ~]# cd /software/kafka_2.11-0.11.0.3/bin/
2. [root@mynode1 bin]# ./kafka-topics.sh --zookeeper mynode3:
   2181,mynode4:2181,mynode5:2181 --list
3. __consumer_offsets
```

经过本次检验命令，发现“firsttopic”成功被删除。

经过以上步骤，可以将 Kafka 中的某个 Topic 管理删除。

## 任务 6.3　测试 Kafka Leader

在 Kafka 中，每个消息队列的分区 Partition 都有副本，这些副本分布在多个 Broker 节点上，每个 Broker 节点管理者一到多个 Topic 的 Partition。如果某些 Broker 节点挂点，那么在 Kafka 集群会将当前 Broker 节点管理的 Partition 按照“副本优先”的机制交由下一个副本所在的 Broker 节点管理，这种机制保证了 Kafka 集群消息的高可用性。

### 6.3.1　Leader 均衡机制

下面通过在 Kafka 中创建一个“mytopic”来说明 Leader 均衡机制的问题。

在 Kafka 任意一台 Broker 节点上执行创建 Topic 命令，这里在 mynode1 中执行命令创建“mytopic”，命令如下：

```
1. [root@mynode1 ~]# cd /software/kafka_2.11-0.11.0.3/bin/
2. [root@mynode1 bin]# ./kafka-topics.sh --zookeeper mynode3:
   2181,mynode4:2181,mynode5:2181 --create --topic mytopic --
   partitions 3 --replication-factor 3
3. Created topic "mytopic".
```

创建了一个 Topic，名称为“mytopic”，这个 Topic 有 3 个 Partition，每个 Partition 有 3 个副本，执行以下命令来查看当前“mytopic”的描述。

```
1. [root@mynode1 ~]# cd /software/kafka_2.11-0.11.0.3/bin/
2. [root@mynode1 bin]# ./kafka-topics.sh --zookeeper mynode3:
   2181,mynode4:2181,mynode5:2181 --describe --topic mytopic
3. Topic:mytopic PartitionCount:3 ReplicationFactor:3 Configs:
4. Topic: mytopic Partition: 0 Leader: 0 Replicas: 0,1,2 Isr: 0,1,2
```

```
Topic: mytopic Partition: 1 Leader: 1 Replicas: 1,2,0 Isr: 1,2,0
Topic: mytopic Partition: 2 Leader: 2 Replicas: 2,0,1 Isr: 2,0,1
```

通过以上查询“mytopic”消息队列的描述，发现在每个 Partition 都有各自的 Replicas，以“mytopic”消息队列中的 Partition 0 为例，当前分区 0 的 Replicas 为 0，1，2，代表当前 Partition 的副本存在于 0 号 Broker、1 号 Broker、2 号 Broker 上。Partition 0 的 Leader 为 0，代表当前 Partition 0 归属 0 号 Broker 节点管理。

当“mytopic”的 Partition 0 所在的 0 号 Broker 节点宕机时，Partition 0 会按照 Replicas 的顺序寻找下一个 Broker 节点来管理，这样就选择了 1 号 Broker 来管理 Partition 0，同样，如果 1 号 Broker 节点宕机之后，Kafka 集群继续按照 Replicas 的顺序继续找到 2 号 Broker 节点来管理当前的 0 号 Partition，直到所有 Kafka 节点都宕机时，Kafka 集群才不能对外提供数据服务。

假设这里的 3 台 Broker 节点挂点了 2 台 Broker 节点，那么最后存活的 Broker 节点负责 Kafka 集群所有 Topic 的分区的读写和存储任务，这样这台节点的负载比较高。这时，如果通过集群监控工具发现其他两台 Broker 节点挂掉，将其他挂掉的两台 Broker 节点重新启动后，Kafka 集群会将当前这台负载较高节点上管理的 Partition 分区自动分散到其他两台重启后的 Broker 节点上，原则是原来两台 Broker 节点管理哪些 Partition，重启之后还是管理哪些 Partition，这个过程避免了将所有的 Kafka 集群 Topic 的读取、存储压力集中到一台 Broker 节点的现象，相当于做了一次负债均衡，这就是所谓的 Leader 均衡机制。

### 6.3.2 测试 Leader 均衡机制

Leader 均衡机制很好地保证了 Kafka 集群的稳定性，避免将 Kafka 中 Topic 的数据集中由少数 Broker 节点处理的现象。下面验证 Kafka 中的 Leader 均衡机制。

①停止两台 Kafka 集群节点。

将 mynode2 节点的 Kafka 进程和 mynode3 节点上的 Kafka 进程停止，执行如下命令：

```
#在 mynode2 上执行如下命令,停止 Broker:
[root@mynode2 ~]# cd /software/kafka_2.11-0.11.0.3/bin/
[root@mynode2 bin]# ./kafka-server-stop.sh


#在 mynode3 上执行如下命令,停止 Broker:
[root@mynode3 ~]# cd /software/kafka_2.11-0.11.0.3/bin/
[root@mynode3 bin]# ./kafka-server-stop.sh
```

经过以上命令的执行，将 Kafka 节点 1 号 Broker 和 2 号 Broker 停止。

②检查“mytopic”的描述信息。

在 mynode1 上执行如下命令：

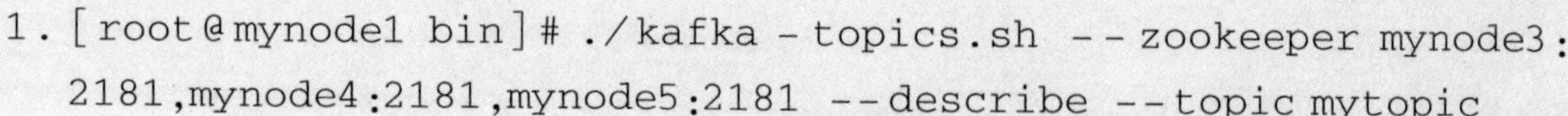

```
[root@mynode1 bin]# ./kafka-topics.sh --zookeeper mynode3:2181,mynode4:2181,mynode5:2181 --describe --topic mytopic
Topic:mytopic PartitionCount:3 ReplicationFactor:3 Configs:
Topic: mytopic Partition: 0 Leader: 0 Replicas: 0,1,2 Isr: 0
Topic: mytopic Partition: 1 Leader: 0 Replicas: 1,2,0 Isr: 0
Topic: mytopic Partition: 2 Leader: 0 Replicas: 2,0,1 Isr: 0
```

经过以上命令的查询，发现 Kafka 集群已经按照“副本优先”的规则将“mytopic”中的每个 Partition 由 0 号 Broker 节点管理。

③重启停掉的两台 Broker 节点。

将 mynode2 节点上的 Kafka 和 mynode3 节点上的 Kafka 重启，执行如下命令：

```
#在 mynode2 节点上重启 Kafka Broker 节点:
[root@mynode2 ~]# cd /software/kafka_2.11-0.11.0.3/bin/
[root@mynode2 bin]# ./kafka-server-start.sh /software/kafka_2.11-0.11.0.3/config/server.properties

#在 mynode3 节点上重启 Kafka Broker 节点:
[root@mynode3 ~]# cd /software/kafka_2.11-0.11.0.3/bin/
[root@mynode3 bin]# ./kafka-server-start.sh /software/kafka_2.11-0.11.0.3/config/server.properties
```

经过以上命令的执行，mynode2、mynode3 节点上的 Kafka 重新启动。

④验证 Leader 均衡机制。

验证“mytopic”中的其他区分是否自动分配给重启后的 Broker 节点管理，在 mynode1 上执行查看“mytopic”的命令：

```
[root@mynode1 ~]# cd /software/kafka_2.11-0.11.0.3/bin/
[root@mynode1 bin]# ./kafka-topics.sh --zookeeper mynode3:2181,mynode4:2181,mynode5:2181 --describe --topic mytopic
Topic:mytopic PartitionCount:3 ReplicationFactor:3 Configs:
Topic: mytopic Partition: 0 Leader: 0 Replicas: 0,1,2 Isr: 0,1,2
Topic: mytopic Partition: 1 Leader: 1 Replicas: 1,2,0 Isr: 0,1,2
Topic: mytopic Partition: 2 Leader: 2 Replicas: 2,0,1 Isr: 0,1,2
```

经过以上查询，发现“mytopic”分区中的各个分区自动均分给每个 Broker 节点管理，Kafka Leader 均衡机制的验证成功。

通过本项目的学习，知道了 Kafka 集群的架构原理、Kafka 应用场景、创建 Kafka Topic 的命令、操作 Kafka Topic 的命令、Kafka 保证数据可靠的机制等。以上这些学习都为 ×× 系统的基础支撑做准备，未来可以使用搭建好的 Kafka 集群为 ×× 系统做数据缓存，使用 Kafka 的操作命令为 ×× 系统的稳定提供更好的数据保障。